U.S. Department
of Transportation
**National Highway
Traffic Safety
Administration**

DOT HS 810 827

September 2007

Review of Technology to Prevent Alcohol-Impaired Crashes (TOPIC)

This document is available to the public from the National Technical Information Service, Springfield, Virginia 22161

This publication is distributed by the U.S. Department of Transportation, National Highway Traffic Safety Administration, in the interest of information exchange. The opinions, findings, and conclusions expressed in this publication are those of the authors and not necessarily those of the Department of Transportation or the National Highway Traffic Safety Administration. The United States Government assumes no liability for its content or use thereof. If trade or manufacturer's names or products are mentioned, it is because they are considered essential to the object of the publication and should not be construed as an endorsement. The United States Government does not endorse products or manufacturers.

REPORT DOCUMENTATION PAGE

Form Approved
OMB No. 0704-0188

Public reporting burden for this collection of information is estimated to average 1 hour per response, including the time for reviewing instructions, searching existing data sources, gathering and maintaining the data needed, and completing and reviewing the collection of information Send comments regarding this burden estimate or any other aspect of this collection of information, including suggestions for reducing this burden, to Washington Headquarters Services, Directorate for Information Operations and Reports, 1215 Jefferson Davis Highway, Suite 1204, Arlington, VA 22202-4302, and to the Office of Management and Budget, Paperwork Reduction Project (0704-0188), Washington, DC 20503

1 AGENCY USE ONLY (Leave blank)	2 REPORT DATE July 2007	3 REPORT TYPE AND DATES COVERED September 2007

4 TITLE AND SUBTITLE **Review of Technology to Prevent Alcohol-Impaired Crashes (TOPIC)**	5 FUNDING NUMBERS

6 AUTHOR(S) John K. Pollard, Eric D. Nadler, Mary D. Stearns[1]	

7 PERFORMING ORGANIZATION NAME(S) AND ADDRESS(ES) U.S. Department of Transportation Research and Innovative Technology Administration John A. Volpe National Transportation Systems Center Advanced Safety Technology Division Cambridge, MA 02142	8 PERFORMING ORGANIZATION REPORT NUMBER DOT HS 810 833

9 SPONSORING/MONITORING AGENCY NAME(S) AND ADDRESS(ES) U.S. Department of Transportation National Highway Traffic Safety Administration	10 SPONSORING/MONITORING AGENCY REPORT NUMBER

11 SUPPLEMENTARY NOTES 1 = Volpe National Transportation Systems Center, RITA, US DOT

12a DISTRIBUTION/AVAILABILITY STATEMENT This document is available to the public through the National Technical Information Service, Springfield, Virginia 22161.	12b DISTRIBUTION CODE

13 ABSTRACT (Maximum 200 words)
This report summarizes the results of an evaluation of vehicular technology alternatives to detect driver blood alcohol concentration and alcohol-impaired driving. Taking an international perspective, this report references relevant literature, incorporates input from stakeholders, and includes a concept of operations to describe how to implement technology-based countermeasures that addresses concerns such as privacy, public acceptance, and legal issues.

14 SUBJECT TERMS Alcohol-impaired driving, alcohol-related fatalities, blood alcohol concentration, breath alcohol ignition interlock, driving under the influence, driving when intoxicated, crash avoidance, collision-avoidance system, ignition interlock device, Safe, Accountable, Flexible, Efficient Transportation Equity Act: A Legacy for Users (SAFETEA-LU), tissue spectroscopy	15 NUMBER OF PAGES 108
	16 PRICE CODE

17 SECURITY CLASSIFICATION OF REPORT Unclassified	18 SECURITY CLASSIFICATION OF THIS PAGE Unclassified	19 SECURITY CLASSIFICATION OF ABSTRACT Unclassified	20 LIMITATION OF ABSTRACT

NSN 7540-01-280-5500

Standard Form 298 (Rev 2-89)
Prescribed by ANSI Std 239-18
298-102

METRIC/ENGLISH CONVERSION FACTORS

ENGLISH TO METRIC

LENGTH (APPROXIMATE)
- 1 inch (in) = 2.5 centimeters (cm)
- 1 foot (ft) = 30 centimeters (cm)
- 1 yard (yd) = 0.9 meter (m)
- 1 mile (mi) = 1.6 kilometers (km)

AREA (APPROXIMATE)
- 1 square inch (sq in, in^2) = 6.5 square centimeters (cm^2)
- 1 square foot (sq ft, ft^2) = 0.09 square meter (m^2)
- 1 square yard (sq yd, yd^2) = 0.8 square meter (m^2)
- 1 square mile (sq mi, mi^2) = 2.6 square kilometers (km^2)
- 1 acre = 0.4 hectare (he) = 4,000 square meters (m^2)

MASS - WEIGHT (APPROXIMATE)
- 1 ounce (oz) = 28 grams (gm)
- 1 pound (lb) = 0.45 kilogram (kg)
- 1 short ton = 2,000 pounds (lb) = 0.9 tonne (t)

VOLUME (APPROXIMATE)
- 1 teaspoon (tsp) = 5 milliliters (ml)
- 1 tablespoon (tbsp) = 15 milliliters (ml)
- 1 fluid ounce (fl oz) = 30 milliliters (ml)
- 1 cup (c) = 0.24 liter (l)
- 1 pint (pt) = 0.47 liter (l)
- 1 quart (qt) = 0.96 liter (l)
- 1 gallon (gal) = 3.8 liters (l)
- 1 cubic foot (cu ft, ft^3) = 0.03 cubic meter (m^3)
- 1 cubic yard (cu yd, yd^3) = 0.76 cubic meter (m^3)

TEMPERATURE (EXACT)
$[(x-32)(5/9)]$ F = y C

METRIC TO ENGLISH

LENGTH (APPROXIMATE)
- 1 millimeter (mm) = 0.04 inch (in)
- 1 centimeter (cm) = 0.4 inch (in)
- 1 meter (m) = 3.3 feet (ft)
- 1 meter (m) = 1.1 yards (yd)
- 1 kilometer (km) = 0.6 mile (mi)

AREA (APPROXIMATE)
- 1 square centimeter (cm^2) = 0.16 square inch (sq in, in^2)
- 1 square meter (m^2) = 1.2 square yards (sq yd, yd^2)
- 1 square kilometer (km^2) = 0.4 square mile (sq mi, mi^2)
- 10,000 square meters (m^2) = 1 hectare (ha) = 2.5 acres

MASS - WEIGHT (APPROXIMATE)
- 1 gram (gm) = 0.036 ounce (oz)
- 1 kilogram (kg) = 2.2 pounds (lb)
- 1 tonne (t) = 1,000 kilograms (kg) = 1.1 short tons

VOLUME (APPROXIMATE)
- 1 milliliter (ml) = 0.03 fluid ounce (fl oz)
- 1 liter (l) = 2.1 pints (pt)
- 1 liter (l) = 1.06 quarts (qt)
- 1 liter (l) = 0.26 gallon (gal)
- 1 cubic meter (m^3) = 36 cubic feet (cu ft, ft^3)
- 1 cubic meter (m^3) = 1.3 cubic yards (cu yd, yd^3)

TEMPERATURE (EXACT)
$[(9/5)y + 32]$ C = X F

QUICK INCH - CENTIMETER LENGTH CONVERSION

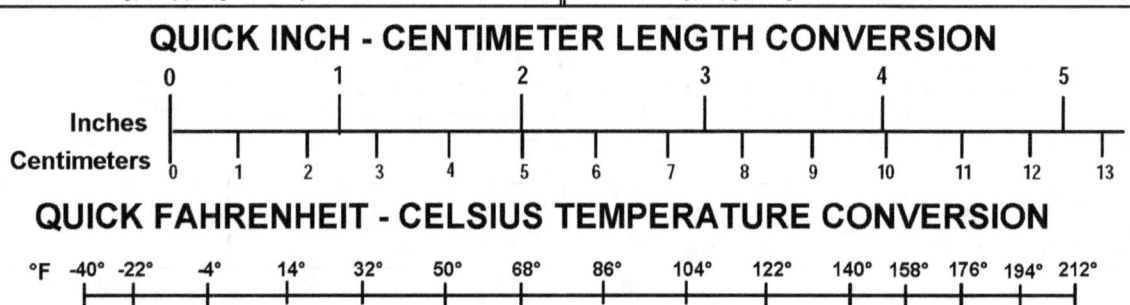

QUICK FAHRENHEIT - CELSIUS TEMPERATURE CONVERSION

°F	-40°	-22°	-4°	14°	32°	50°	68°	86°	104°	122°	140°	158°	176°	194°	212°
°C	-40°	-30°	-20°	-10°	0°	10°	20°	30°	40°	50°	60°	70°	80°	90°	100°

For more exact and or other conversion factors, see NIST Miscellaneous Publication 286, Units of Weights and Measures. Price $2.50 SD Catalog No. C13 10286 Updated 6/17/98

Acknowledgements

The authors of this report are *John K. Pollard, Eric D. Nadler*, and *Mary D. Stearns,* all of the Volpe Center. The authors acknowledge the technical contribution and support of many individuals. Special appreciation is due to *Mike Perel*, the NHTSA program manager; *John Hinch*, director, Office of Human-Vehicle Performance Research; *Joseph Kanianthra*, associate administrator for research; and *Julie Barker and Eric Traube*, NHTSA, for their support and technical guidance. *Wiel Janssen*, of TNO, and *Nic Ward*, of the University of Minnesota, provided outstanding assessments of the technical issues associated with monitoring driver impairment. *Michael Regan, Max Cameron*, and *Michael Lenne,* all of Monash University, provided valuable perspectives on the Australian efforts to reduce alcohol-related driver impairment as well as an assessment of the effort on this topic in South Asia. *Susan Partyka, Chou-Lin Chen*, and *Rajesh Subramanian*, NCSA at NHTSA provided guidance on statistical resources. *Susan Kelly*, Gosnold Counseling Center, provided insights about treatment programs for DUI offenders. Geoffrey Collier, NHTSA; Jim Hedlund, a private consultant; and Sue Ferguson, Ferguson International LLC reviewed and made very helpful comments on an earlier draft of this report. The authors had the opportunity to make three presentations to NHTSA and found the feedback and guidance provided during these meetings to be useful in steering the research program.

The authors also acknowledge the following Volpe Center staff who contributed to many aspects of this examination: *Stephen Popkin*, chief, Human Factors Division, for his support; *Art Flores* and *Ed Conde* for consultation on issues related to the measurement of alcohol impairment; and *Andrea Sparko* and *Amy Ricci* for their assistance with data analysis queries.

LIST OF ACRONYMS

ABS	Antilock Brake System
ACC	Adaptive Cruise Control
ADHD	Attention Deficit Hyperactivity Disorder
AMS	Alcohol Monitoring Systems, Inc.
APS	Administrative Program Suspension
AWAKE	System for Effective Assessment of Driver Vigilance and Warning According to Traffic Risk Estimation
BAC	Blood Alcohol Concentration
BAIID	Breath Alcohol Ignition Interlock Device
BrAC	Breath Alcohol Concentration
CANbus	Controller Area Network (local area network widely used in vehicles)
CDC	Centers for Disease Control and Prevention
DALY	Disability-Adjusted Life-Year
DRE	Drug Recognition Expert
DREAM	Driver-Related Evaluation and Monitoring
DRUID	DRiving Under the Influence of Drugs
DUI	Driving Under the Influence
DWI	Driving When Intoxicated
DWS	Driving While Suspended
EEG	Electroencephalograph
EOG	Electro-oculograph
ESC	Electronic Stability Control
EU	European Union
FARS	Fatal Accident Reporting System
FCW	Forward Collision Warning
fNIR	Functional Near Infrared
FOT	Field Operational Test
GDP	Gross Domestic Product
GES	General Estimates System
GPS	Global Positioning System

GSR	Galvanic Skin Response
HGN	Horizontal Gaze Nystagmus
HIPPA	Health Information Privacy Protection Act
ICC	Intelligent Cruise Control
IID	Ignition Interlock Device
IMU	Integrated Monitoring Unit
ISF	Interstitial Fluid
ISA	Intelligent Speed Adaptation
ISO	International Standards Organization
ITS	Intelligent Transportation Systems
IVI	Intelligent Vehicle Initiative
LAN	Local Area Network
LDWS	Lane Departure Warning System
LED	Light Emitting Diode
LIDAR	Light Detection and Ranging (infrared radar)
MHF	Motorförarnas Helnykterförbund (Swedish Abstaining Motorists' Association)
NCSA	National Center For Statistics And Analysis
NHTSA	National Highway Traffic Safety Administration
NIAAA	National Institute on Alcohol Abuse and Alcoholism
NIR	Near-Infrared
NSDUH	National Survey on Drug Use and Health
PDA	Personal Digital Assistant
PIRE	Pacific Institute for Research and Evaluation
QDOT	Quantum-dot LEDs
RMS	Root Mean Square
RT	Response Time
SAFETEA-LU	Safe, Accountable, Flexible, Efficient Transportation Equity Act: A Legacy For Users
SAVE	System For The Effective Assessment of the Driver State and Vehicle Control in Emergency Situations
SAVE-IT	Safety Vehicle Using Adaptive Interface Technology
SCRAM	Secure Continuous Alcohol Monitor

SD	Standard Deviation
SDLP	Standard Deviation of Lane Position
SENSATION	Advanced Sensor Development for Attention, Stress, Vigilance, and Sleep/Wakefulness Monitoring
TDLS	Tunable-Diode Laser Spectroscope
TOPIC	Technology to Prevent Alcohol-Impaired Crashes
TNO	The Netherlands Organization for Applied Scientific Research
USDOT	U.S. Department of Transportation
VMT	Vehicle Miles Traveled
WM	Working Memory
WrisTAS	Wrist Transdermal Alcohol Sensor

TABLE OF CONTENTS

1. INTRODUCTION.. 1-1
   - 1.1 ISSUES 1-1
   - 1.2 GOAL 1-1
   - 1.3 APPROACH 1-1
   - 1.4 OVERVIEW OF REPORT 1-2
2. CHARACTERISTICS OF ALCOHOL-RELATED CRASHES................................ 2-1
3. POTENTIAL FOR CRASH REDUCTION WITH TECHNOLOGY TO PREVENT ALCOHOL IMPAIRED CRASHES.. 3-1
4. TECHNOLOGIES IN USE 4-1
   - 4.1 BREATH ALCOHOL IGNITION INTERLOCK DEVICES (BAIIDs) FOR DUI OFFENDERS... 4-1
     - 4.1.1 Description 4-1
     - 4.1.2 Performance and Limitations 4-2
     - 4.1.3 Per Se and Behavioral Impairment 4-5
     - 4.1.4 Accident Statistics for Interlock Users..4-8
     - 4.1.5 Rate of Interlock Use...4-11
   - 4.2 OTHER NEAR-TERM TECHNOLOGIES... 4-11
     - 4.2.1 Description 4-11
     - 4.2.2 Performance Limitations 4-13
5. TECHNOLOGIES UNDER DEVELOPMENT 5-1
   - 5.1 TISSUE SPECTROSCOPY 5-1
     - 5.1.1 Description 5-1
     - 5.1.2 Validation Experiments 5-3
     - 5.1.3 Limitations 5-3
     - 5.1.4 Alternative Implementations 5-4
   - 5.2 NEW TECHNOLOGIES TO DETECT ALCOHOL VAPOR 5-5
     - 5.2.1 Descriptions 5-5
     - 5.2.2 Applications 5-6
     - 5.2.3 Limitations 5-7
   - 5.3 ENVIRONMENTAL MONITORS 5-8
   - 5.4 VEHICLE-BASED IMPAIRMENT MONITORS 5-9
     - 5.4.1 Use of Physiological and Vehicle Sensors to Detect Alcohol Impairment...5-9
     - 5.4.2 Use of Physiological and Vehicle Sensors to Distinguish Alcohol From Other Impairment Sources 5-12
     - 5.4.3 Vehicle-Based Impairment Detection Using Multiple Sensors.............5-14
6. COMPARISON OF TECHNOLOGIES IN USE AND UNDER DEVELOPMENT .. 6-1
   - 6.1 DESCRIPTION OF TECHNOLOGIES AND RANKING BY ACCURACY..................... 6-2

	6.2 COSTS	6-5
	6.3 LATENCY	6-7
	6.4 USABILITY	6-8
	6.5 TECHNICAL RISK	6-9

7. CROSS-CUTTING IMPLEMENTATION ISSUES 7-1
    7.1 ACTIVE VERSUS PASSIVE METHODS 7-1
    7.2 PRE-START VERSUS POST-START TESTING 7-1
    7.3 CHOICE OF THRESHOLD TO TRIGGER COUNTERMEASURES 7-2
    7.4 PRIVACY ISSUES AND CIRCUMVENTION 7-2
    7.5 VEHICLE TELEMATICS 7-3
    7.6 TARGET USER GROUPS 7-3
    7.7 COUNTERMEASURE ALTERNATIVES 7-4
        7.7.1 Alcohol Warning (Qualitative or Binary) 7-5
        7.7.2 BAC Warnings (Quantitative) 7-6
        7.7.3 Impairment Warnings 7-6
        7.7.4 Performance and Incentive Feedback 7-7
        7.7.5 Crash Avoidance Warnings 7-8
        7.7.6 Crash-Avoidance Technologies 7-9
        7.7.7 Potential for Distraction from In-Vehicle Warnings 7-10
        7.7.8 Acceptability of Incentives and Driving Performance Feedback 7-11

8. A CONCEPT OF OPERATIONS FOR A TECHNOLOGY UNDER DEVELOPMENT: PRIMARY INTERLOCK USING TISSUE SPECTROSCOPY 8-1
    8.1 OBJECTIVES AND TARGET USERS 8-1
    8.2 PHYSICAL CHARACTERISTICS AND PERFORMANCE SPECIFICATIONS 8-1
        8.2.1 Accuracy 8-2
        8.2.2 Set Point 8-2
        8.2.3 Latency 8-3
        8.2.4 Form Factor 8-3
        8.2.5 User Friendliness 8-4
        8.2.6 Feedback 8-4
        8.2.7 Retesting Intervals 8-4
        8.2.8 Environment 8-5
        8.2.9 Reliability and Maintenance Burden 8-5
        8.2.10 Verification That Tissue Sample Tested Belongs to Driver ... 8-6
        8.2.11 Anti-Circumvention Features 8-6
        8.2.12 Vehicle Interface 8-6
        8.2.13 Economic Constraints 8-7
        8.2.14 Technology Bundling 8-9

9. CONCLUSIONS 9-1
    9.1 TECHNOLOGIES IN USE: SECONDARY INTERLOCKS 9-1
    9.2 TECHNOLOGIES UNDER DEVELOPMENT: PRIMARY INTERLOCKS 9-1

    9.3   Technologies Under Development: Vehicle-Based Impairment Monitors 9-2

10. RECOMMENDATIONS FOR TECHNOLOGIES IN USE 10-1

11. RESEARCH RECOMMENDATIONS FOR TECHNOLOGIES UNDER DEVELOPMENT 11-1

    11.1 NIR Tissue Spectroscopy 11-1
    11.2 Role of Warning Devices 11-1
    11.3 A General Behavioral Baseline for Impairment Detection 11-2
        11.3.1   Background and Research Strategy 11-2
        11.3.2   Risks 11-5
    11.4 Countermeasure Research 11-6
        11.4.1   Impairment and Incentive Displays 11-6

12. REFERENCES 12-1

LIST OF TABLES

Table 2-1 Alcohol-related crashes by number of vehicles involved and severity, 2004 2-1
Table 3-1 Alcohol-related fatality rate per 100 million VMT, 2002-2004 3-1
Table 3-2 Crash victims by person type, 2004 [5] 3-2
Table 3-3 Vehicles involved in crashes by type and severity in 2004 3-3
Table 5-1 Average Performance of Project SAVE Classification Regimes 5-15
Table 6-1 Comparison matrix for primary interlock applications 6-2
Table 11-1 Alcohol involvement in crash scenarios 11-5

LIST OF FIGURES

Figure 4-1 Comparison of estimates of BAC from BrAC analysis with true BAC 4-3
Figure 4-2 Alcohol and drug co-morbidity 4-6
Figure 4-3 Crash free survival rates for drivers receiving an ignition interlock device (IID) order/restriction versus comparison group of DWS/DWI offenders not using interlocks 4-9
Figure 4-4 Crash-free survival rates for IID users (all offenders in administrative program) versus comparison group of DUI offenders not using interlocks 4-10
Figure 4-5 Survival curves (Kaplan-Meier method) for single-vehicle nighttime crashes 4-11
Figure 5-1 Comparison of estimates of BAC from NIR spectroscopy with true BAC 5-3
Figure 5-2 Linearity and sensitivity characteristics of perovskite oxide sensor 5-6

Executive Summary

PURPOSE

The National Highway Traffic Safety Administration's Office of Human-Vehicle Performance Research tasked the Volpe National Transportation Systems Center (Volpe Center) of the U.S. Department of Transportation's Research and Innovative Technology Administration to identify current and emerging vehicle-based technologies that can detect driver blood alcohol concentration (BAC) and monitor driver impairment due to alcohol. Detection technologies have the potential to prevent death and injury by monitoring BAC and/or driving performance for signs of impairment, and if necessary either prevent ignition or take other actions to prevent a crash.

As part of the SAFETEA-LU legislation, the Secretary of Transportation was directed to conduct a study on reducing the incidence of alcohol-related motor vehicle crashes and fatalities through research on advanced vehicle-based alcohol detection systems, including an assessment of the practicability and effectiveness of such systems. In support of this mandate, this report on Technology to Prevent Alcohol Impaired Crashes (TOPIC) assesses the capability of existing and anticipated technologies to detect and prevent alcohol-impaired driving. It also includes a concept of operations to describe how to implement technology-based countermeasures while addressing concerns such as privacy, public acceptance, and legal issues.

SCOPE

The incidence of BAC involvement in fatal crashes has dropped during the last 25 years, but the rate of improvement has leveled off during the past 10 years.[1] In 2004, there were 12,677 fatalities in crashes attributable to a driver with a BAC ≥ .08 grams per deciliter and they account for 76 percent of the alcohol-related fatalities. It is estimated that universal use of secondary interlocks[1] by all DUI offenders as well as all commercial vehicle operators and drivers under age 21 would have reduced motor vehicle fatalities in 2004 by 10 percent, from 42,636 to 38,373. Universal adoption of primary interlocks would have produced a 30-percent decrease in motor vehicle fatalities in 2004.

TECHNOLOGIES IN USE

---

[1] A primary interlock is one intended for all users regardless of prior driving history. A secondary interlock is one imposed as a result of a DUI offense.

The breath-alcohol ignition interlock device (BAIID) is an aftermarket product hardwired into the ignition circuit of a vehicle that prevents starting until a breath sample has been given, analyzed for ethanol content, and found to be below programmed limits. Currently, about a third of repeat-DUI offenders are using interlocks, along with a very small fraction of first offenders. Collectively, there are only about 100,000 units in use, as compared with more than one million DUI arrests per year.

BAIIDs have been found to reduce DUI recidivism by 40 percent to 90 percent in various studies. However, crash rates for interlock users are higher than for nonusers, because the latter have their licenses revoked and tend to drive less and with particular effort to avoid police attention. Best available data indicates that the crash rates of the interlock users are essentially equal to those of average drivers. The low rate of use of BAIIDs is mostly the result of institutional factors, rather than shortcomings in the technology. However, technology improvements over the next decade are likely to decrease costs and inconvenience to users by extending the interval between visits to have the BAIID serviced.

Solid-state breath alcohol monitors are sold as screening devices and have been proposed for primary interlocks. They lack the accuracy and ethanol-specificity of fuel cells, but have substantial advantages in terms of size, cost, and power consumption, especially for installation in a cell phone or a key fob. Recently developed solid-state detectors are claimed to have much better accuracy and specificity than the tin-oxide cells (Taguchi cells, named after the inventor) found in most screening devices in current use. Some of the prototypes for primary interlocks developed in Sweden use these new technologies, but details are proprietary, as are data on the accuracy and reliability of these devices. The Swedish government is considering making them mandatory in new vehicles in a few years, and they are being installed in test fleets now.

TECHNOLOGIES UNDER DEVELOPMENT

TISSUE SPECTROSCOPY

Spectroscopes are devices that measure the proportion of a beam of light that is absorbed or reflected by a sample at various wavelengths. The concentration of ethanol in tissue changes its absorption of near-infrared (NIR) light at certain wavelengths. This phenomenon allows estimation of BAC by measuring how much light has been absorbed at particular wavelengths from a beam of NIR reflected from the tissue of the subject. Infrared light easily penetrates several millimeters of tissue; hence the reflected signal reveals information about the tissue to that depth. This makes NIR reflectance spectroscopy relatively insensitive to contaminants on the surface of the skin. Because the reflected spectrum is affected by many other chemicals present in the skin, the estimation relies on a complex statistical process called a partial-least-squares model.

The accuracy of a statistical estimation process depends on the quantity and quality of the input data, which is a function of the number of different wavelengths that are measured and the number of times each is sampled. Data quality is affected by physical properties of the detector, such as bandwidth, noise, linearity, and stability. Achieving narrow bandwidths at low cost is particularly challenging. Reducing the size, cost, and measurement time of the tissue-spectrometer while maintaining data quality will require a substantial effort in technology development, testing, and refinement. There are also physiological questions that must be resolved. The soft, thin skin on the underside of the forearm works well for reflectance spectroscopy. Little is known about the reflectance characteristics of the thicker, tougher skin of the palms and fingers, or perfusion rates in various parts of the hand, or the effects of the bones that lie close to the skin.

Initial published data comparing estimates of BAC made with tissue spectroscopy against true BAC show excellent correlation. These results represent levels of accuracy, sensitivity, and specificity to ethanol that are far superior to other known methods of measuring alcohol impairment that do not involve extraction of bodily fluids. Testing of a prototype by the Bernalillo County, New Mexico, Sheriff's Department will begin in the autumn of 2007.

NEW TECHNOLOGIES TO DETECT ALCOHOL VAPOR

Among the methods of ethanol vapor detection that have recently been invented are the following:
- tunable-diode laser spectroscopes;
- carbon nanotubes that exhibit changes in optical or acoustic properties proportional to the concentration of ethanol vapor flowing through them;
- nano-crystalline perovskite oxides doped with strontium that selectively catalyze the oxidation of ethanol and measure the energy released; and
- solid-polymer-electrolyte sensors.

Any of these methods might be developed as a device to detect ethanol vapor in vehicles and notify authorities of the fact. This approach might be used to enhance the effectiveness of police checkpoints, but vapor detection is not recommended for use with interlocks because the relation between ethanol concentration in the driver's blood and ethanol concentration in the air in a vehicle varies far too much.

Some of these technologies are also being used for body-worn ethanol monitors. They find application both in clinical studies and in the enforcement of court-ordered abstinence. While it is technically feasible to extend their uses to include interlocks, they would compare unfavorably with other technologies in terms of cost, accuracy, and inconvenience to users.

VEHICLE-BASED IMPAIRMENT DETECTION

Countermeasures that detect impaired driving through objective behavioral measures could decrease crash risk beyond what is possible through direct alcohol detection. Progress has been made in defining physiological measures that have potential for in-vehicle use. Ocular, gaze, and eye-movement measures have demonstrated sensitivity to alcohol impairment, but their implementation in vehicles remains a challenging problem, especially in sunlight. Other physiological measures require head-worn sensors that render them unattractive for general use.

Some progress has been achieved in the detection of alcohol impairment through driving-performance measures, as exemplified in the European Union's Project System for the Effective

Assessment of the Driver State and Vehicle Control in Emergency Situations (SAVE), which employed a multiplicity of sensors feeding a neural network. In the SAVE experiments, the variables included:

- Eye blink;
- Eyelid closure;
- Steering wheel grip;
- Mean lane position (relative to right lane marking);
- SD of lane position;
- SD of steering wheel position;
- Mean speed;
- SD of speed; and
- Time to lane crossing.

However, using state of the art technology in the form of individualized neural-network detectors, SAVE correctly detected alcohol (BAC = .05 g/dL) in only 78 percent of trials and generated false alarms in 8 percent of the trials. Additional progress is required before this approach can be considered for practical use.

The primary benefit of behavioral definitions is that they will detect impairment caused by BAC levels less than the per se limit, such as impairment resulting from low levels of alcohol combined with fatigue or other factors. However, some amount of driving is required before detection occurs, so unlike ignition interlocks, they cannot completely prevent impaired driving. Further, it is important to individualize baselines when assessing the effects of alcohol, drugs, and medicines on driving behavior. The "natural" or unimpaired behavior of the driver should be known before additional effects can be estimated. The individualized "signature" of alcohol-influenced driving may include indicators that have no relevance to crash risk. For this reason, a general baseline comprised of behaviors that result in increased crash risk must be combined with the alcohol signature to identify alcohol impaired driving.

CONCEPT OF OPERATIONS

A Concept of Operations is presented that describes an implementation of primary interlocks based on tissue spectroscopy. It assumes that technology will be invented to permit large reductions in size and cost for this approach and that will allow the sensors to be embedded in devices drivers normally handle, such as key fobs or steering wheels. BAC readings will be taken in a few seconds and securely transmitted to a computer in the vehicle. If a BAC in excess of the per se limit is detected, the vehicle's computer may prevent starting or restrict the use of the vehicle by limiting speed, flashing the hazard warning lights, etc.

Implementation of this concept requires the development of detectors and control components of very high reliability and inherent invulnerability to circumvention. They must operate for the life of the vehicle without periodic visits to service centers to compensate for the shortcomings of current detectors and to protect them from tampering.

CONCLUSIONS AND RECOMMENDATIONS

Secondary interlocks have demonstrated effectiveness in reducing DUI recidivism, and have the potential to reduce the crash rates of DUI offenders to approximate those of other drivers. Fewer than 8 percent of offenders use them. Many issues contribute to their low rate of use ranging from insufficient judicial awareness of their potential to concerns about cost. Addressing these institutional issues has great potential to reduce alcohol-related crashes in the near term.

There is no technology ready for near-term use as a primary interlock. Tissue spectroscopy has the most promising characteristics, but based on Volpe Center estimates must be reduced by three orders of magnitude in size, one order of magnitude in cost, about an order of magnitude in measurement time and re-designed to work on palms or fingers.

Current technologies are highly vulnerable to circumvention without an infrastructure that permits devices to report circumvention attempts to authorities that can impose appropriate sanctions. An interlock must be developed that is inherently invulnerable to circumvention through:

- Secure integration with the vehicle's engine control computer
- Inclusion of additional test features to verify that the sample tested is from the driver, and
- Capability to perform accurately and reliably throughout the life of the vehicle.

Although there are other approaches to the detection of alcohol-impairment through physiological or behavioral measures, they appear inferior to tissue spectroscopy in terms of accuracy, measurement time, and/or convenience.

Priorities for near-term research to reduce alcohol-related crashes include:

- Establishing the credibility of NIR-tissue spectroscopy for wider application in interlocks through the development and field-testing of a portable evidential device.
- Determining whether, and how, tissue spectroscopy can be used on palms or finger tips (as opposed to the underside of the forearm used in current instruments).
- Evaluating the efficacy of impairment warnings and incentive displays and identifying ways to display warning information that will not increase impairment through distraction.

1. INTRODUCTION

1.1 ISSUES

In 2004, the National Highway Traffic Safety Administration reported that there were 16,694 deaths and 248,000 people injured as a result of alcohol-related motor vehicle crashes.[2, 3] Alcohol-related motor vehicle fatalities account for 39 percent of all motor-vehicle-related deaths. The NHTSA Administrator has stated that this fatality rate is a national concern during her 2006 testimony to Congress, referring to the provisions in the Safe, Accountable, Flexible, Efficient Transportation Equity Act: A Legacy For Users (SAFETEA-LU) legislation to provide increased funding to reduce impaired driving.

Fatalities and injuries due to motor vehicle crashes create particularly heavy losses to society when expressed as disability-adjusted life-years (DALYs), because the motor vehicle crash victims tend to be young. (DALYs tabulate the number of years lost to premature death and disabling injuries.) The public health community is increasingly aware of the losses world-wide due to vehicle accidents and forecasts that vehicle accidents will move from the ninth cause of DALYS in 1990 to the third leading cause by 2020. [4] The 16 to 20 and 20 to 24 age groups have the highest fatality rate per 100,000 -- more than double the rate for the overall population, [5] with a substantial proportion of these crashes being alcohol-related.

1.2 GOAL

As part of the SAFETEA-LU legislation, the Secretary of Transportation was directed to conduct a study on the potential for reducing the incidence of alcohol-related motor vehicle crashes and fatalities through advanced vehicle-based alcohol detection systems. Also included is an assessment of the practicability and effectiveness of such systems.[6] NHTSA's Office of Human-Vehicle Performance Research has tasked the Volpe National Transportation Systems Center (Volpe Center) to assess the potential for vehicle-based technologies to prevent alcohol-impaired crashes. The purpose of this TOPIC report is to identify vehicle-based technologies capable of detecting and preventing alcohol impaired driving. Research data will provide input to a report required to be submitted in 2007 to Congress.

1.3 APPROACH

Impairment detection technologies have the potential to prevent serious crashes by stopping impaired drivers from starting or operating vehicles. Technology can detect driver BAC and lock out the driver or monitor driving performance for signs of impairment. If driver impairment is evident, the system can warn the driver and or impose any of a range of measures to mitigate risk.. This report assesses the ability of technologies, existing and anticipated, to detect driver impairment from alcohol and identifies international state-of-the-art vehicle-based technology options to prevent alcohol-impaired automotive crashes. This analysis was carried out with the support of the Intelligent Vehicle Initiative (IVI) program, created in 1998 as part of the Department of Transportation ITS program. The IVI program focuses on the collision warning system as an effective tool to reduce the number of accidents by providing effective and timely warnings to drivers.

The Volpe Center team interviewed stakeholders and interested parties and reviewed research results to assess the capability of vehicular technologies to reduce alcohol-impaired driving. The team acquired academic expertise both to review post-1995 literature and to identify international sources of expertise for vehicle-based alcohol impairment detection.

The research team collaborated with European experts to acquire first-hand information about the results of the EU projects: System for the Effective Assessment of the Driver State and Vehicle Control in Emergency Situations (SAVE) and System for Effective Assessment of Driver Vigilance and Warning According to Traffic Risk Estimation (AWAKE), the plans and intent of the EU project Advanced Sensor Development for Attention, Stress, Vigilance and Sleep/Wakefulness Monitoring (SENSATION), and the status of relevant research by European institutes on vehicle-based alcohol impairment identification and countermeasures. In addition, the European collaborator conducted personal interviews with six major European stakeholders representing technology developers such as original equipment manufacturers and suppliers and EU government agencies about vehicle-based alcohol impairment identification and countermeasures.

1.4 OVERVIEW OF REPORT

This report describes the most effective means of measuring alcohol impairment and the strategies for implementing technology-based countermeasures. The report estimates the potential for crash reductions as a result of introducing TOPIC in relation to fatalities and injuries avoided; describes the strengths and weaknesses of near-term approaches, including breath alcohol ignition interlock devices (BAIID) for DUI offenders; evaluates current and emerging crash avoidance technologies; identifies cross-cutting implementation issues likely to accompany the introduction of these technologies; identifies research needs; and provides a concept of operations using promising technologies.

2. CHARACTERISTICS OF ALCOHOL-RELATED CRASHES

Alcohol-related crashes are distinguished by their severity, overrepresentation of recidivist DUI offenders, a disproportionate occurrence at certain times of day, and overrepresentation of certain age groups.

Table 2-1 shows that alcohol-related crashes are more likely to result in loss of life and to involve single vehicles. Almost two-fifths, 39 percent, of the alcohol-related crashes in 2004 resulted in a fatality.[2] Almost half, 47 percent, of the alcohol-related crashes in 2004 involved a single vehicle compared to 28 percent involving multiple vehicles.

Table 2-1 Alcohol-related crashes by number of vehicles involved and severity, 2004 [2]

Alcohol-Related Single Vehicle		Alcohol-Related Multiple Vehicle		Alcohol-Related Total	
Number	Percent	Number	Percent	Number	Percent
Fatal Crashes					
10,307	47	4,661	28	14,968	39
Injury Crashes					
92,000	16	76,000	6	168,000	9
Property Damage Only					
138,000	11	110,000	4	247,000	6

Alcohol-impaired drivers in fatal crashes are more likely to have been speeding. In 2003, 41 percent of the alcohol impaired (BAC=.08+) drivers in fatal crashes were speeding compared to the 14 percent of drivers in fatal crashes with zero BACs. [8]

Alcohol-impaired drivers are less likely to have valid driver licenses at the time of the crash. In 2004, 9 out of 10 drivers in fatal crashes with BAC = 0 had valid licenses compared to 76 percent of drivers with BACs of .08 to .14, and 73 percent of those with BACs ≥ 0.15.[9]

Alcohol-related crashes occur more often at certain times of day and days of the week. More alcohol-related crashes occur at night. From 9 p.m. to 6 a.m., using three-hour intervals, the proportion of crashes that are alcohol-related ranges from 60 percent to 76 percent. [2] The proportion of injury-related crashes involving alcohol during the same time period ranges from 22 percent to 39 percent and property damage-only crashes ranges from 14 percent to 27 percent.

---

[2] Alcohol-related refers to fatalities that occur in a crash involving at least one driver, motorcycle operator, pedestrian, or pedalcyclist with a BAC of .01 g/dL or greater.[7]

In 2004, there were 1,433,382, DUI[3] arrests for alcohol and/or drugs, including 1,014,000 DUI arrests for alcohol. Using the base of 198, 888, 912 licensed drivers in 2004, it is estimated that the DUI for alcohol arrest rate was one half of one percent of the licensed drivers. However, retrospective self-reports from surveys suggest that many more people admit to driving while under the influence of alcohol. One-fifth to one-fourth of drivers surveyed admitted to driving after drinking at least once during the prior year.[11] Other research estimates that on average a drunk driver will drive while impaired between 80 and 2,000 times for every time the driver is apprehended, depending on the enforcement in the locality.[12] The low risk of apprehension may contribute to why people drive while impaired.

The National Institute on Alcohol Abuse and Alcoholism's National Epidemiologic Survey on Alcohol and Related Conditions examined the change in the rate of self-reported driving-after-drinking from 1991–1992 to 2001–2002. The prevalence of self-reported driving after drinking was 2.9 percent in 2001– 2002, corresponding to approximately 6 million adults in the United States. This rate was about three-quarters of the rate observed in 1991–1992 (3.7%), reflecting a 22-percent decline.

Using survey data, it is estimated that there were 906 million unenforced DUI incidents in 2001, which is a decrease from the estimated 967 million incidents in 1996[12, 13][4] This 6-percent decrease, or 1.2 percent/year, may parallel an increasing public awareness of the risks of driving after drinking. It is estimated that there are 91 million trips annually in which drivers have BACs of .08 or greater.[14] Comparing the incidence of DUI arrests for alcohol with the estimated number of alcohol impaired trips suggests that drivers have a 1 percent chance of being arrested for DUI for each trip when they drive over the legal limit (contributing to the number of trips driven by legally impaired drivers are the estimated 17.6 million adult Americans classified as alcoholics in 2002[6], most of whom have driver's licenses).

People arrested for DUI offenses have poorer driving histories prior to their DUI arrest than the general population.[15] Their poorer driving may bring them to the attention of law enforcement. Driving skill may mediate the impairing effect of alcohol on driving. Research shows that people with poorer driving skills when not alcohol impaired demonstrate more impaired driving[5] in simulation when dosed to the BAC=.08 level on within-lane deviation, controlling for alcohol use history.[16] This study helps to explain why there are individual differences in response to the same amount of alcohol.

Recidivist DUI offenders (prior DUI within the past three years) contributed to 8 percent of the fatal crashes in 2004.[9] NHTSA estimates that about one-third of all drivers arrested or convicted of DUI offenses have previous DUI convictions.[17] Earlier Canadian data shows that among drunk drivers responsible for fatal crashes, one-third have previous DUI convictions at some point in their driving careers.[15] Other research estimates that 20 percent to 28 percent of

---

[3] Driving under the influence —Driving or operating a motor vehicle or common carrier while mentally or physically impaired as the result of consuming an alcoholic beverage or using a drug or narcotic.[10]
[4] The number of vehicle trips in 2001 is estimated at 243,191,417,148, based on the National Household Travel Survey. Using this number, .003 percent of these trips were unenforced DUI incidents. (http://nhts.ornl.gov/2001/index.shtml)
[5] Driving skill was indicated as within-lane deviation. Alcohol use history was controlled.

first-time DUI offenders repeat the DUI offense.[18] It is difficult to calculate the incidence of recidivism after a DUI arrest because common terms are not used and records are kept for a limited duration.[19]

Research on the contribution of recidivism to crashes has identified the following characteristics: repeat offenders are more likely than non-DUI drivers to have high BACs (BAC=.15+) when arrested, more fatal motor vehicle crashes, and more hit and runs with pedestrians when arrested[18]; repeat offenders have poorer driving records and may have poor driving skills; they also differ from first-time offenders on psychopathology and psychiatric distress measures. [20]

Social trends in drinking patterns may foster alcohol-impaired driving. In our interviews, Centers for Disease Control and Prevention personnel commented on the extent of, and recent increase in, binge drinking.[21] Binge drinking is common in most segments of society in the U.S.[6] Most people who binge drink are not classified as alcohol-dependent and the CDC expects some of the younger ones to "age out" of the behavior.[23]

Nationally, in 2001, 16 percent of adults reported binge drinking.[24] Binge drinkers are 14 times more likely to drive while impaired by alcohol compared with non-binge drinkers.[23] Although the rates of binge-drinking episodes were highest among those 18 to 25 years old, 69 percent of binge-drinking episodes during the study period occurred among those 26 or older.[23] Almost half, 47 percent, of binge-drinking episodes occurred among otherwise moderate drinkers and 73 percent of all binge drinkers were moderate drinkers.

The rate of binge drinking increases with age from 18 to 21, but decreases with 22-year-olds, regardless of college enrollment status. The highest rate of binge drinking among underage people was among full-time college students and other 21-year-olds.[25] About 90 percent of the alcohol consumed by youth under age 21 in the United States is consumed in binges.[23]

---

[6] Binge drinking is defined as drinking five or more drinks on the same occasion on at least one day in the past 30 days.[22]

3. POTENTIAL FOR CRASH REDUCTION WITH TECHNOLOGY TO PREVENT ALCOHOL-IMPAIRED CRASHES

This section describes the methodology for estimating the impact of widespread use of TOPIC and the number of alcohol-related crashes TOPIC might prevent. These estimates provide a baseline to assess the potential impact of a universal adoption of TOPIC.

Although the incidence of alcohol-related fatalities has declined over time, the rate of decline has slowed and developments in technology, such as TOPIC, may offer a way to prompt more dramatic decreases. The proportion of high-BAC involvement in fatal crashes has dropped from three-fifths to two-fifths of the crashes during the last 25 years, but the rate has leveled off to around two-fifths of the crashes during the past 10 years.[14] NHTSA expresses the incidence as alcohol-related fatalities per 100 million vehicle miles traveled (VMT) to reflect exposure.[26] Table 3-1 Alcohol-related fatality rate per 100 million VMT, 2002-2004[28] shows the incidence of fatalities and alcohol-related fatalities in recent years. A longer view shows the rate of fatalities per 100 million VMT with driver BAC=.08+ dropped from 1.46 in 1984 to 0.64 in 1994 and 0.43 in 2004.[27]

Table 3-1 Alcohol-related fatality rate per 100 million VMT, 2002-2004[28]

Year	Total fatalities per 100M VMT	Fatalities per 100M VMT, BAC=.01+	Fatalities per 100M VMT, BAC=.08+
2002	1.50	0.62	0.53
2003	1.48	0.59	0.45
2004	1.44	0.56	0.43

Table 3-2 shows the incidence of motor vehicle fatalities and injuries in the United States in 2004. Drivers and passengers account for almost four-fifths (78%) of the motor vehicle fatalities and almost all (93%) of the motor vehicle crash injuries.[5]

Table 3-2 Crash victims by person type, 2004[5]

Crash Victims	Fatalities		Injuries	
	Number	Percent	Number	Percent
Occupants	33,134	78	2,594,000	93
Drivers	23,063	54	1,782,000	64
Passengers	9,991	23	811,000	29
Unknown	80	0	1,000	0
Motorcycle riders[7]	4,008	9	76,000	3
Nonmotorists[8]	5,494	13	118,000	4
Pedestrians	4,641	11	68,000	2
Pedalcyclists	725	2	41,000	2
Other/unknown	128	0	9,000	0
Total*	**42,636**	**100%**	**2,788,000**	**100%**

* Aggregates may not equal subcategories due to rounding error.

Table 3-3 describes all police-reported crashes by vehicle type and severity. More than four-fifths of the fatalities (82%) occurred with passenger cars and light trucks. Almost all the injuries and property losses (95%) occurred with passenger cars and light trucks.

---

[7] Alcohol use by motorcycle drivers appears to be a significant phenomenon and it is likely that a subset of technologies to prevent alcohol-impaired driving could be adapted for use by motorcycle riders.
[8] Non-motorists are sometimes referred to as non-occupants. This category includes pedestrians and cyclists.

Table 3-3 Vehicles involved in crashes by type and severity in 2004 [2]

Vehicle Type	Crash Severity						Total
	Fatalities		Injuries		Property Damage-Only		
	Number	Percent	Number	Percent	Number	Percent	
Passenger car	25,507	44	1,990,000	58	4,216,000	56	6,232,000
Light truck	22,337	38	1,246,000	37	2,886,000	39	4,154,000
Large truck	4,862	8	87,000	3	324,000	4	416,000
Bus	275	1	13,000	0	39,000	1	52,000
Other/unknown	635	1	9000	0	10,000	1	20,000
Motorcycles	4,100	7	70,000	2	13,000	0	88,000
Total*	**58,414**	**100**	**3,415,000**	**100**	**7,489,000**	**100**	**10,962,000**

* Aggregates may not equal subcategories due to rounding error.

To estimate the impact of the introduction and use of TOPIC for reducing motor vehicle fatalities, it is necessary to identify the number of drivers involved in fatal crashes who were alcohol impaired. Of the 42,636 traffic fatalities in 2004, there were 16,694 alcohol-related fatalities and 14,409 of these involved someone with a BAC $\geq$ .08, including non-drivers. NHTSA uses the term *alcohol-related* for crashes in which someone involved, whether the driver, passenger, or a nonoccupant, has measurable blood alcohol (BAC $\geq$ .01).

The number of drivers with BACs $\geq$ .08 was 8,256 or 49 percent of the drivers in crashes resulting in alcohol-related traffic fatalities.[9] In 2004 there were an estimated 12,677 fatalities attributable to drunk driving including passengers, the drunk driver, pedestrians, and other drivers.[9]

There is a clarification needed when discussing estimates of drunk driving. BAC results are not available for all drivers and nonoccupants involved in fatal crashes, and the imputed numbers rely on use of estimations for the missing data. For example, States differ in the extensiveness of the BAC testing, ranging from a low of 8 percent to a high of 82 percent.[29]

In his presentation to the International Technology Symposium: A Nation Without Drunk Driving in 2006 in Albuquerque, Fell [9] extrapolated how the impact of widespread adoption of primary as well as secondary interlocks would decrease the incidence of fatalities due to drunk driving. If interlocks, BAIIDs, were installed in the vehicles operated by 100 percent of new and repeat DUI offenders, in commercial vehicles, as well as in all vehicles driven by drivers under age 21, it is estimated that crash deaths resulting from drunk drivers would decrease 25 to 40 percent, preventing an estimated 3,000 to 5,000 fatalities of the 12,677 fatalities in 2004 attributable to drunk driving.

Subsequent chapters in this report describe potential new technologies which might be suitable as primary interlocks. If primary interlocks are developed that are accurate, reliable, durable, invulnerable to circumvention, and installed in all vehicles, they could potentially eliminate all drunk driving at BAC ≥ 0.08. The authors of this document estimate that this would eliminate 30 percent of the traffic fatalities or all 12,677 of the 42,636 fatalities that occurred in 2004.

4. TECHNOLOGIES IN USE

4.1 BREATH ALCOHOL IGNITION INTERLOCK DEVICES (BAIIDs) FOR DUI OFFENDERS

4.1.1 Description

The interlocks in current use are secondary interlocks, i.e., secondary to apprehension for a DUI offense. Most States impose interlocks only on repeat offenders, although there are a few jurisdictions in which their use is an option for first offenders. The BAIID is an aftermarket product hardwired into the ignition circuit of a vehicle that prevents the vehicle from starting until a breath sample has been given, analyzed for ethanol content, and found to be below programmed limits.

For proprietary reasons, the interlock manufacturers are reluctant to release their sales figures and there are no official estimates of the number of interlocks in use. However, in the course of our interviews with the chief executives of all of the U.S. manufacturers, we heard estimates of 85,000 to 100,000 units in use in 2006. These estimates were provided on a "not for attribution" basis.

The BAIID hardware consists of a handheld sensor-and-display unit together with an under-dash unit that contains the interface to the vehicle's ignition and power circuits. Nearly all units now in service contain fuel-cell ethanol sensors, as well as sensors for breath temperature, pressure and/or air flow. A microprocessor controller performs the following functions:

1. Each time a driver attempts to start, the controller first turns on the heater in the fuel cell and delays further action until the proper operating temperature has been reached. In very cold conditions, this may take as much as 3 minutes, but 30 seconds is typical in mild weather.

2. The unit then signals the driver to blow a sample. Accurate estimation of BAC requires the air sample to be from deep in the lungs, so the driver must take a deep breath and blow long and hard. Based on signals from the pressure and flow-rate sensors, the controller limits its analysis of the ethanol concentrations to the last portion of the sample. Blowing a sample with acceptable characteristics in terms of pressure, volume and/or flow rate requires training. Some units include a microphone and demand that the driver hum while blowing. Without training and practice in blowing an acceptable sample, it is difficult for a sober individual to substitute for a drinking driver. Failures to blow an acceptable sample are logged.

3. If an acceptable sample is blown and found to contain less than the programmed limit for ethanol – usually .02 or .025 BAC among DUI drivers in the U.S. – the vehicle can then be started normally.

4. If the sample exceeds the limit, the ignition is locked out for some period of time and the date, time, and ethanol concentration are logged. After some period of time – typically 5 to 30 minutes – the controller signals that another sample may be given.

5. Once driving has begun, at random intervals ranging from a few minutes to nearly an hour, the controller signals the driver to blow another sample, called a "rolling retest." This sample must be given within a few minutes of the signal, or a failure will be recorded. Drivers are cautioned to pull over out of traffic to perform the retest, but many ignore this advice. At least one serious accident is known to have occurred because of distraction associated with a retest. This feature serves both to prevent drinking while driving and circumvention. Failure to complete a rolling retest may also trigger a requirement to visit a service center within 5 days; otherwise the vehicle will be disabled until an authorized service technician appears to clear the interlock (at considerable expense to the vehicle owner).
6. Disconnection and/or bypassing of the interlock is detected and recorded.
7. A data log of all events of interest (sample failures, denied starts, missed retests, circumvention attempts, etc.) is maintained in non-volatile memory and may be downloaded and erased only at an authorized interlock service center owning the required equipment.

Participants in interlock programs are required to visit a service center periodically (usually monthly) to have their data logs downloaded, fuel cell sensors replaced with recalibrated units, and the complete system checked for proper operation. If a vehicle is more than seven days late for a periodic visit, its ignition will be disabled.

The interlock service providers collect and assemble the data according to the requirements of the jurisdiction. Most commonly, the data (usually with only failures, circumvention attempts, etc.) are posted on a secure website. Some jurisdictions want the data sent by email or fax, and some want complete reports of every piece of information captured by the interlocks.

4.1.2 Performance and Limitations

The BAIID is a mature technology with performance characteristics that are generally adequate to sharply reduce DUI recidivism among the offenders required to use it. There are inherent limitations in breath-alcohol testing as a means of estimating true blood alcohol concentration, and the relation of BAC to impairment varies somewhat among individuals. However, these are of little consequence in the offender-interlock application, because its intent is to prevent driving after any drinking. Discrimination between BAC levels of .07 versus .08 is far more challenging.

Breath alcohol concentration (BrAC) analyzers were developed in the 1950s and have become the main proof of intoxication in prosecutions. Earlier devices for the evidential market involved multistage wet chemistry and photometry. Currently, infrared spectrometers are considered the most accurate technology and dominate the evidential market. Some include a fuel cell to provide two independent tests from each driver. The NHTSA specification for the error (standard deviation) in measurements is .0042 g/dL BrAC for both evidential instruments and interlocks. The manufacturers of the evidential instruments publish an error specification on the order of .003 g/dL BrAC.

While these instruments can measure breathe alcohol quite accurately, that is not the same as measuring BAC. The ratio of BrAC to BAC, known as the "partition ratio," varies between 1:1,900 and 1:2,400. In the United States, it is defined by statute to be 2,100, but other nations have selected other values. Within measurements on the same individual, the partition ratio has been shown to vary with body temperature, vigorous exercise, presence of alcohol in the mouth, and whether BAC is rising or falling. To obtain readings that will stand up in court, police officers wait at least 15 minutes to ensure that mouth alcohol has dissipated and keep the driver seated in a controlled-temperature environment. Breath temperature is tested to ensure that it is within the narrow limits that allow a valid reading.

Because partition ratios vary, to avoid "reasonable doubt" in court, BrAC analyzers are biased to estimate BAC lower than its true value on average, as shown in Figure 4-1.

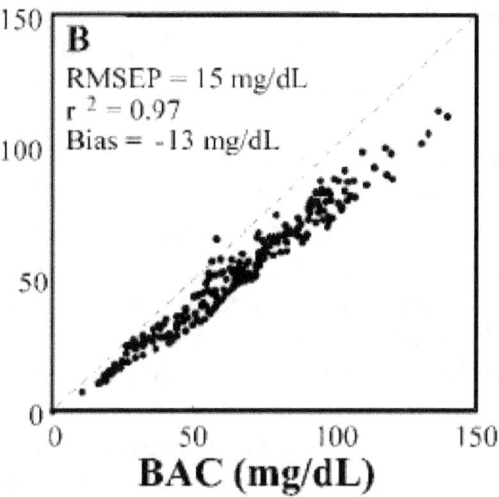

Figure 4-1 Comparison of estimates of BAC from BrAC analysis with true BAC[9]

The current specification for accuracy in BAIIDs devices is that they must lock out at least 90 percent of the time (18 out of 20 trials) when the alcohol concentration in the test sample exceeds the set point (normally .025 BrAC) by .01 BrAC. At extreme temperatures or under other conditions of stress, the allowable deviation from set point increases to .02 BrAC.

Most interlock manufacturers use **fuel-cell sensors** today. This technology is fairly rugged and ethanol specific, but the fuel cell must be warmed up to breath temperature to meet the accuracy specification. This requires a heater assembly and significant energy use for heating; this is not a problem for a device that is hardwired to a vehicle, but is a major barrier to the use of fuel cells in wireless devices like key fobs.

For applications requiring small size and low battery drain, **solid-state sensors** are used to measure breath alcohol. When freshly calibrated, they can be almost as accurate as fuel cells, but they show considerable drift over time. Furthermore, they respond to several volatile organic compounds other than ethanol. Since their use for enforcement purposes has not been sanctioned, there is little public-domain data regarding their accuracy. Contamination is also a problem. Recent research suggests substantially improved accuracy and specificity may be obtained by replacing the current tin-oxide sensor with one constructed from perovskite crystals doped with strontium, but currently no complete monitors with this technology are available for testing. Recently developed gallium arsenide detectors are now being tested in Sweden.

The low interlock set point prohibits any driving after drinking. Because the standard prediction error in current interlocks amounts to at least .015 BAC from true BAC, in most U.S. jurisdictions, offender interlocks are programmed to trigger at .02 or .025 BAC -- far below the per se limit of .08 BAC, but comfortably larger than the error margin to prevent false positives. It might be possible for some individuals to consume one drink and still be allowed to start driving shortly thereafter, but this is by no means certain.

Data on the reliability of current sensors is unavailable. No States are known to actively monitor data regarding BAIID failures -- even where there are laws or regulations that require the interlock service providers to report such data. There is anecdotal evidence that substantial numbers of interlock users have complained about erroneous readings, but the complaints have apparently not been investigated. An analyst who has examined large numbers of interlock data records notes that there are suspiciously high numbers of instances in which BrAC readings taken only a few minutes apart show substantial differences. These could be caused by hardware problems or by persons other than the driver providing the samples. Instances often occur in morning daylight hours with an initial reading of .04 or .05 followed by a passing reading. This is suggestive of a driver who was drinking the night before, and asks another family member to blow a passing sample so that the person can drive to work.

Current fuel-cell sensors are known to exhibit some drift in response -- about 1 percent of the reading per month, requiring frequent replacement of the sensor with a freshly calibrated unit. Service providers would like to increase the recalibration interval to reduce costs and inconvenience, but the feasibility is still debatable. A related issue is quality control in recalibration operations. Some vendors perform all recalibration in a central laboratory, while others do it in their local service centers. Anecdotal reports suggest that some of the local personnel do not follow appropriate procedures.

Contamination of sensors is another problem, but again there is no available data about its incidence. These failures may cause the unit to issue a lockout warning, meaning that the vehicle must be returned to a service center within seven days for a repair. There is no charge for such a visit, but it is an obvious inconvenience to the users.

Under the current certification process for interlocks, manufacturers are required to have 30 separate tests performed on a given product, as detailed in *Model Specifications for Breath Alcohol Ignition Interlock Devices.*[30] However, the States require only that vendors submit letters attesting that products have passed all of the tests performed by independent labs. This is in contrast to the more reliable practice of requiring copies of the independent test lab reports, which must then be examined by the staff of the entity requiring the certification. It has been alleged that some interlocks currently in use do not actually meet all of the requirements. Their manufacturers have had various samples tested by different labs. Some samples passed some criteria at one lab, some at another, and some at a third. Collectively, the samples passed all of the tests at least once at a lab, but no single sample passed all of the tests. Furthermore, it is alleged that there have been numerous instances in which the design of products has changed, but model numbers are not retained. Therefore, there remains some doubt about the accuracy, reliability, and durability of current BAIIDs.

Current interlocks are designed to enforce a zero-tolerance policy for DUI offenders driving after drinking which, in part, is a way to account for the error band surrounding a BAIID measurement. For comparison, evidential grade fuel cell monitors have a standard error of about +/- .015. It is safe to assume that the error for interlock sensors is higher, because they are built to sell at much lower prices and because the interval between calibrations is much longer.

4.1.3 Per Se and Behavioral Impairment

Per se laws have provided a valuable legal framework that facilitates the identification and conviction of dangerously alcohol-impaired drivers. They are based on research that determined the relative risk associated with particular BAC levels.[31] Law enforcement countermeasures are more effective when based on objective criteria such as BAC than on subjective criteria. The same reasoning applies to BAIID technology in that it relies on a per se BAC threshold for ignition. The potential for objective behavioral measures of impairment is that they could supplement the per se definition where research has indicated conditions under which the per se definition leads to errors in misclassifying risky driving as unimpaired. Evidence of these misclassification errors is reviewed in this section.

In their review of the information-processing effects of alcohol alone and in combination with other drugs, Kerr and Hindmarch refer to "the large variation in response to alcohol found not only between individuals but also within an individual on different occasions" ([32], p. 2). In their conclusion, they remark, "the most striking feature of the literature on the effects of small doses[10] of alcohol on cognitive function and psychomotor performance is the variability in the results that are reported" (p. 5). More recently, summarizing individual differences in response to alcohol, Harrison and Fillmore noted that, "even when participants receive a standardized dose of alcohol and attain the same blood alcohol level... some individuals display a large degree of impairment while others display little or no impairment" ([16], p.883). It follows from the extent of this variability that use of a per se definition will misclassify some individuals who are capable of driving safely as impaired, and more importantly, will misclassify some who are not capable of driving safely as unimpaired.

The comorbidity of alcohol and other impairment sources is another weakness of per se definitions because they can create a dissociation of BAC and risk. Evidence of a dissociation of BAC

---

[10] Kerr and Hindmarch define BAC < .10 as a small or low dose.

and degraded vehicle behavior (which implies increased risk) can be found in studies on poly-drug impairment and in studies of the conjunction of fatigue and alcohol impairment. In their analysis of blood test results reported by coroners or medical examiners, Terhune et al. [33] found that 11 percent of fatal crashes involved alcohol-drug comorbidity (see Figure 4-2). Kerr and Hindmarch [32] reviewed evidence for additive or "super-additive" effects of some benzodiazepines and antidepressants, and alcohol. Antihistamines, narcotic analgesics, some anti-infective agents and nonprescription cold medications can increase the effects of alcohol. These effects were found using laboratory tasks. Robbe [34] studied the effect of cannabis alone and in combination with alcohol in a highway road-tracking task, where the subject was to maintain a constant speed and center lane position. Alcohol (BAC = .04) in combination with the lowest cannabis dose studied increased lane position variability to a level equivalent to BAC = .08 and, in combination with a higher cannabis dose, increased lane position variability to levels equivalent to BAC = .14 as established in separate research.[35] A later study reported that the same alcohol dose alone (BAC = .10) achieved a similar lane position variability effect.[36] Lamers and Ramaekers[37] tested subjects' visual search for traffic at intersections during actual city driving. The subjects wore an eye tracker to provide evidence of visual search. Neither alcohol (BAC = .05, the applicable per se limit) nor cannabis significantly affected visual search for traffic at intersections, but the combination of these drugs significantly reduced visual search by 3 percent.

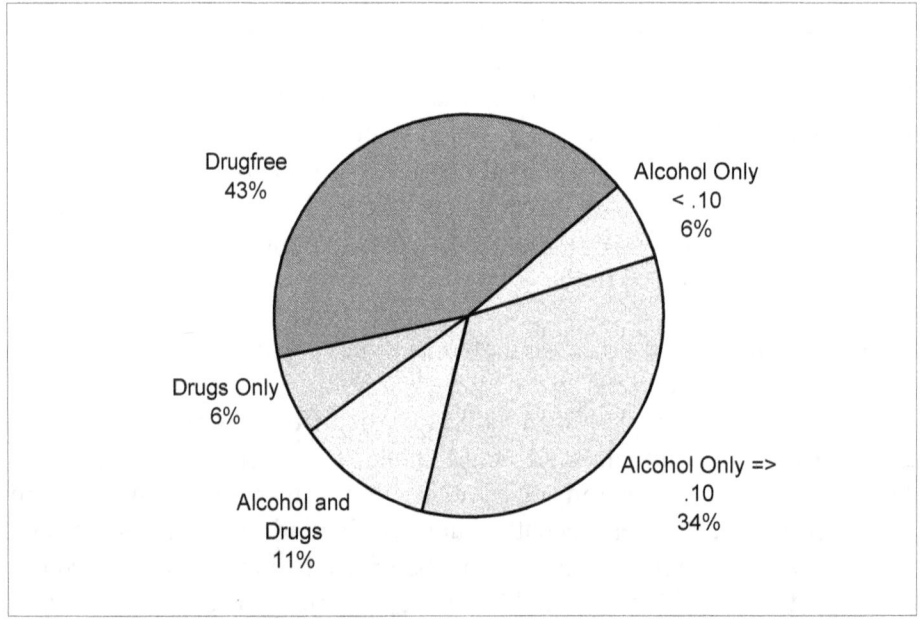

Figure 4-2 Alcohol and drug co-morbidity

Sources of impairment other than drugs also can combine with alcohol to increase some of its deleterious effects on driving. For example, the effects of fatigue and limited visibility contribute to an overall 141-percent elevated risk of a crash for driving at night. With particular drivers and conditions, almost half of this increase has been shown to result from alcohol impairment.[38] Keall, Frith, and Patterson[39] found that even a low dose of alcohol significantly increased the night driving risk for teenage drivers. Driving simulator research indicates that alcohol (BAC =

.04) significantly increased the effects of partial sleep deprivation during simulated driving and in addition increased breaking reaction time, steering deviation, and microsleeps.[40] Other studies have found that the combination of alcohol (BAC = .04) and restricted sleep increased lane drifting during simulated driving.[41, 42]

These studies suggest that lower BACs combined with other impairment sources can have a similar effect on driving as BACs that are above the per se limit. The prevalence of fatal crashes where alcohol and another drug were found underestimates the comorbidity weakness of per se definitions. It does not include fatalities associated with combinations of alcohol and non-drug factors that combine to cause levels of driving impairment that are at least as high as the average impairment found at the per se limit.

Differences in impairment also exist based on whether alcohol is being absorbed or eliminated from the body. More impairment is found at the same BAC during the absorption phase than during the elimination phase. Known as the Mellanby Effect, this difference has been demonstrated by Wang for proprioceptive responses[43] and by Grattan-Miscio and Vogel-Sprott for working memory.[44]

Grattan-Miscio and Vogel-Sprott also found that working memory decrements return to normal levels in stages as BAC declines.[44] In particular, increases in working memory response times due to alcohol return to unimpaired levels sooner (at a higher BAC) than accuracy. These effects were found with BACs between .05 and .08. In our interview, Fillmore suggested that the subjective awareness of impairment abates prior to effects on performance. If so, subjective awareness could represent another instance where descending BACs show stage-like effects. Thus impairment does not appear to a unitary, all or none phenomenon that occurs beyond a particular per se limit. To entirely avoid both misses and false positive errors, different per se limits would apparently be needed during alcohol absorption and elimination, an impractical alternative.

Cortical (electroencephalograph or EEG) effects have been shown over the first 35 minutes following alcohol consumption for BACs ascending to .03. This evidence suggests that a non-zero per se limit is unlikely to represent a boundary that distinguishes BAC levels that can affect human behavior (i.e., EEG) and those that cannot (cf. [45]).

In summary, the per se definition of impairment is limited in several ways. It does not distinguish BACs that have no physiological (EEG) and behavioral (working memory, proprioception) effects on humans from those that have an effect. It implies that impairment is a unitary phenomenon that occurs after BAC reaches a specified threshold, whereas the actual threshold can differ for ascending and descending BACs, and the impairment can recover in a stage-like manner, not all at once. Comorbidity of alcohol at sub-threshold levels with drugs and other impairment sources such as fatigue, and the variability evident in the effects of alcohol when BAC is constant, produce a dissociation of BAC and crash risk that would limit any nonzero BAC criterion. Because of these limitations countermeasures that rely solely on a per se BAC limit will fail to prevent alcohol-impaired driving that exhibits as much or greater risk than driving at a BAC beyond the per se limit.

4.1.4 Accident Statistics for Interlock Users

BAIIDs have been demonstrated to be quite effective in reducing DUI recidivism. Various studies report rates of DUI offenses among interlock users that are 40 to 90 percent lower than those of control groups; usually repeat-DUI offenders in the same jurisdiction who are not interlock users. However, reductions in DUI recidivism do not translate directly into crash-rate reductions, as described in the studies below.

Ideally, exposure-weighted crash rates for interlock users should be compared with corresponding rates for a carefully matched group of non-users. No such study has been done, because of the enormous methodological difficulties (especially with respect to randomization of subject assignments in a judicial setting), privacy issues, and expense. Longitudinal studies of interlock users would also be helpful, but only a few have been performed, and to our knowledge, none include data from the period prior to interlock use. In all of these studies, crash rates are expressed in terms of time, rather than exposure. Only one study has included random assignment of offenders to interlock use or the control group, and it did not track crash rates. In all of the other analyses, offenders, judges, or hearing officers decide who is given an interlock, which results in a substantial selection-bias problem. It is reasonable to suppose that individuals who need to drive often and can afford to do so get interlocks; those who are poor or do not need to drive much accept license revocation. The latter often continue to do some driving, and they continue to be apprehended for DUI. Their annual VMT is thought to be much less than before revocation, but actual data are lacking.

Two studies that shed light on the crash-risk effects of interlock use are summarized as follows:

An Evaluation of the Effectiveness of Ignition Interlock in California[46, 47]

This analysis by the R&D Section of the California DMV contains comparisons of six different categories of DUI offender with controls. All of the comparisons are expressed in terms of Cox regression models, which plot the proportion of drivers with no subsequent DUI conviction or crash (vertical axis) against the elapsed time since their previous DUI offense. These are commonly known as "survival models." Some of the comparisons address the offenders sentenced by courts and compare crash rates against sentence type. Others are based on data from interlock service providers and apply to drivers who actually installed an interlock under the California administrative program.

Among drivers convicted of both DUI and DWS (driving while suspended), one group (n=6,742) received either a court order to install an ignition interlock or a license restriction prohibiting use of a vehicle without an interlock. This group had crash rates 24 percent lower than the controls. For similar drivers receiving the interlock-installation order only (n=1,691), the crash rate was 42 percent lower than controls, as shown in Figure 4-3.

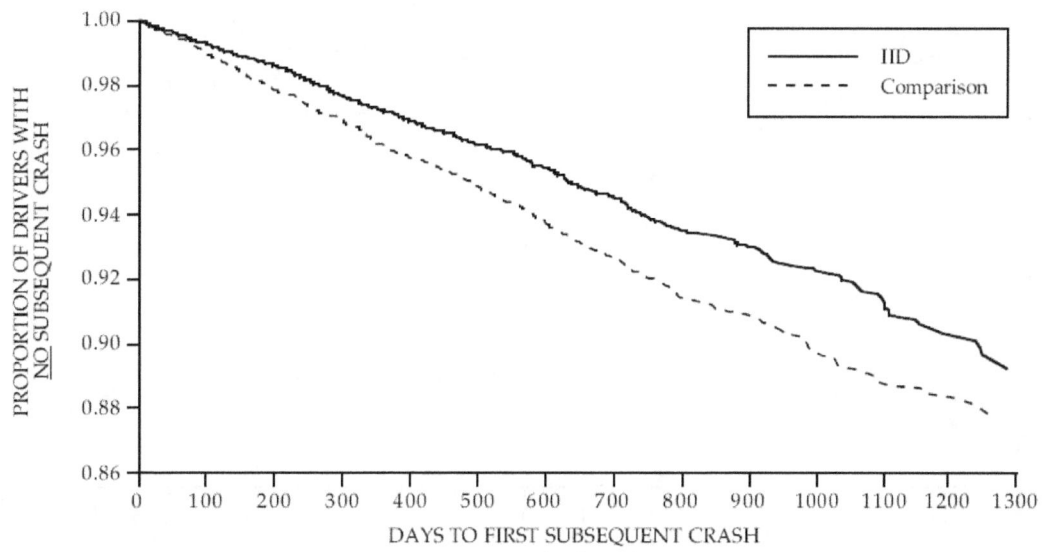

Figure 4-3 Crash-free survival rates for drivers receiving ignition interlock devices (IIDs) order/restriction versus comparison group of DWS/DWI offenders not using interlocks

An analysis of the effects of judicially-mandated interlock use showed that interlocks had no significant effect on subsequent crash rates for first-DUI offenders (n=1,227), but that among second-DUI offenders (n=5,416), the crash rate was 19 percent lower for interlock users. These conclusions are based on analyses of data regarding actions of the courts and administrative actions of the DMV – not the actual use of interlocks by the offenders. By checking records of interlock providers, the authors discovered that only about 20 percent of the California drivers ordered to install interlocks actually did so. Their hypothesis is that most of the reduction in daily crash-risk for the group receiving such orders is the result of reduced VMT and/or a more conservative driving behavior that lessens the likelihood of a crash.

Other studies examined drivers who actually had interlocks installed under the administrative program between January 2000 and January 2003. The risk of a subsequent crash was 84 percent higher for the interlock group than for a control group whose licenses remained revoked, as shown in Figure 4-4 [47]

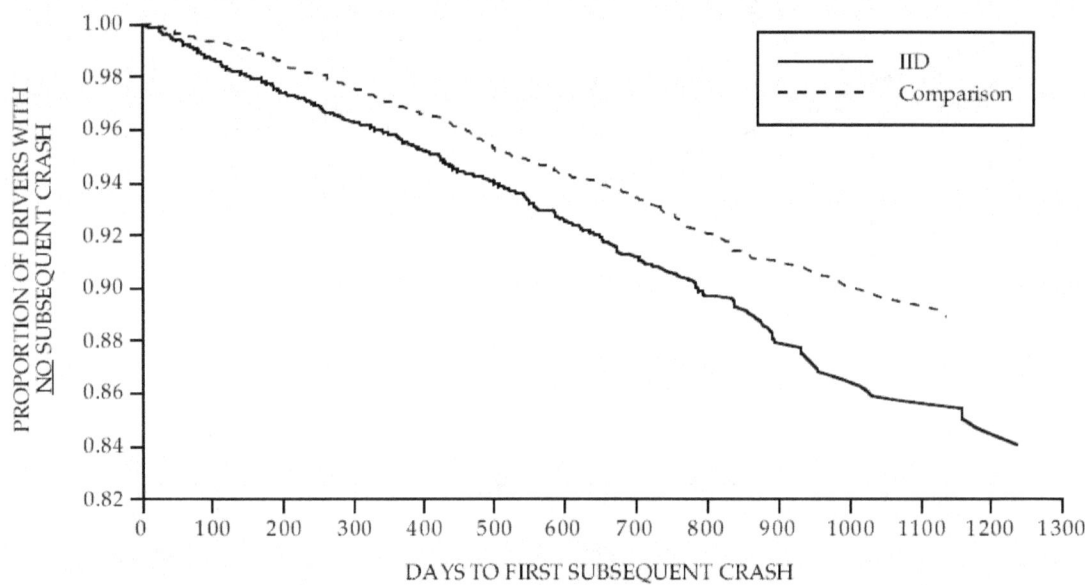

Figure 4-4 Crash-free survival rates for IID Users (all offenders in administrative program) versus comparison group of DUI Offenders not using interlocks

For repeat offenders who installed interlocks under the administrative program (n=600) daily crash rates were 130 percent higher than for the controls. The presumed explanation for the higher crash rates of interlock users is simply that they drove much more than non-users. This study did not collect odometer readings and was not able to report data on miles driven. Thus the California study did not find evidence that interlock use reduced daily crash rates compared with offenders whose licenses were revoked.

Quebec Alcohol Ignition Interlock Program: Impact on Recidivism and Crashes [48]

This study considered data from the crashes of a group of 42,563 drivers in Quebec who were sentenced for impaired driving between December 1, 1997, and January 26, 2001. Of these drivers, 9,896 first offenders and 1,050 repeat offenders were allowed to have interlocks installed. The other drivers' licenses were revoked. Separate analyses were conducted using the entire set of crashes and using the subset of single-vehicle nighttime crashes in order to focus on the ones most likely to involve alcohol. Cox models were developed with data stratified both by first and repeat offenders and by elapsed time since the offence that brought them into the study.

The risk of crashes was higher for drivers with interlocks under all conditions. Survival curves for single-vehicle nighttime crashes were plotted separately for first and repeat offenders. As shown in Figure 4-5 these demonstrate small, but statistically significant differences over the three-year study with higher rates associated with interlock use. The authors agreed with Morse and Elliot's conclusion that interlock users are involved in more crashes because they drive more than drivers with licenses revoked.[49]

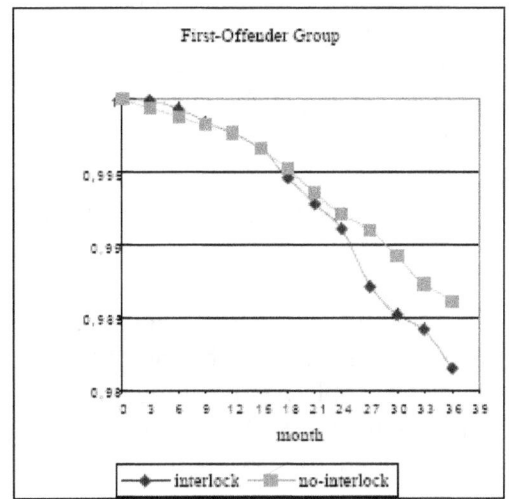 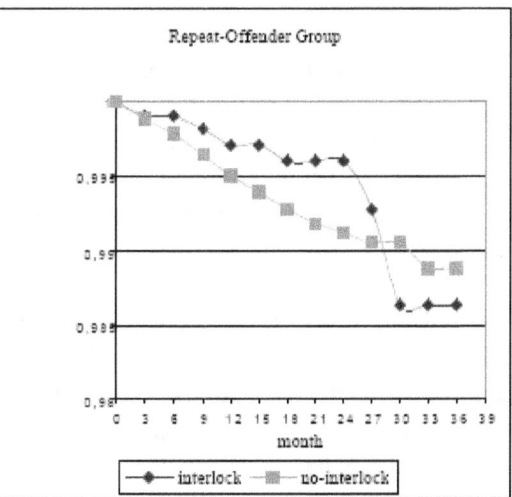

Figure 4-5 Survival curves (Kaplan-Meier method) for single-vehicle nighttime crashes

4.1.5 Rate of Interlock Use

Offender and institutional acceptance of interlock technology appears to constitute an impediment in its increased use. Interlock vendors estimated the total number of units currently in use in the United States as 80,000 to 100,000 as compared with estimates of 300,000 to 400,000 repeat arrests for DWI each year. The relatively low rate of usage is the result of institutional factors outside the scope of this report.

In New Mexico, for example, the recidivism rate of first-DUI offenders with interlocks was found to be less than one-half that of nonusers.[50] However, survey research in California has indicated that there is stronger hostility toward interlocks among first offenders than repeaters, possibly because the former have not yet recognized the extent of their alcohol problems. More information is needed to guide implementation of interlock programs for first offenders.

According to the interlock providers that we interviewed, numerous experts have identified the need to educate judges about how interlocks work. They told us that prosecutors say that they cannot give lectures during trials and so the task is passed on to the interlock providers. Some providers assign representatives to visit judges one-on-one to explain the technology. However, some judges do not grant time for such visits, and others do not trust what they hear from salesmen. One approach suggested by two Virginia attorneys is that the NHTSA Regional Administrators arrange presentations for judges at the annual or biennial educational conferences that occur in most States.[51]

4.2 OTHER NEAR-TERM TECHNOLOGIES

4.2.1 Description

Although they are no longer much used in BAIIDs, solid-state alcohol Taguchi cell monitors continue to be sold in substantial numbers as screening devices and have been proposed for primary interlocks. Even though they lack the accuracy and ethanol-specificity of fuel cells, they have substantial advantages in terms of size, cost, and power consumption. Tin oxide has been used as the active component in most of these devices, but other chemistries are now being researched and tested in Europe and Asia. Most of these developments are proprietary. For installation in a cell phone or key fob, the small size, low cost, and low power use of solid-state detectors are essential.

Many low-cost ($25 to $200) screeners are being marketed. The physical sizes of components have been reduced to the point that it is feasible to incorporate screeners into other products, such as cell phones. LG Electronics has announced in 2006 that it would begin selling a cell phone with a screener (the LG 4100) in the United States, although no U.S. dealers appear to be selling it at this writing. Sales in Korea are reported to have exceeded 200,000 units.

When a BAC over .08 is detected, the LG-4100 displays a "wavy car" icon, and also blocks calls to a preprogrammed list of numbers.

While these proliferating screeners are not examples of in-vehicle technology, their development seems to account in part for the perception that primary interlocks could be made using their detectors. Interest in primary interlocks is greatest in Sweden, where legislation requiring interlocks in all new vehicles by 2012 is under consideration. Saab and Volvo have developed prototypes of primary interlocks. The Saab model is built into a key fob and uses a solid-state sensor. Its trigger level is programmable to the per se limit of the country in which the vehicle is to be sold. The estimated retail price is about $300, but marketing plans remain uncertain. Model year 2008 was rumored.

Prototypes of the Saab Alcokey

Volvo is testing various prototypes, apparently including both solid-state and fuel cell detectors. One version is embedded in the in the seat belt buckle. In one concept, this device would warn by means of a dash indicator if any alcohol was detected and enforce seat belt use. It would also disable the ignition if the statutory BAC limit (.02) were exceeded.

4.2.2 Performance Limitations

Data on the accuracy and reliability of these devices is unavailable. It is known that the start of mass production has been delayed several times due to technical problems. Nevertheless, the Swedish government is considering making them mandatory in new vehicles in a few years, and many Swedish units of local government are already buying them for fleets.

Volvo BAIID in a seat belt buckle concept

The effort to establish widespread use of primary interlocks in Sweden is being led by the Motorförarnas Helnykterförbund (MHF, or the Abstaining Motorists' Association). This organization maintains an extensive Web site describing its activities and news about interlock experiments. Some pages are in English, and all issues of its newsletter, *Alcolock News*, are available in translation for download. See http://www.mhf.se/

MHF tests interlocks at its laboratory in Tibro. Its newsletter contains reports that some of the screening devices marketed in Sweden are terrible and that some fail to detect high levels of alcohol. *Alkolåsnytt* (#1, 2006) contains evaluations by MHF of nine interlocks now on the market (in Swedish only). Two were rated unsatisfactory for any application – both brands unknown in the United States. All the products that are in use in North America received the highest rating: Transport Quality. Interestingly, the selling prices about approximately double what they are in the United States. Recommended service intervals ranged from one to four times per year.

Alcolock News includes reports on pilot tests in government fleets – some of which are negative. Even government workers routinely hot-wired them in one test fleet because virtually all of the units were generating false positives. Fleet managers were angry with the government for failing to conduct laboratory tests before starting in-service field trials. There are numerous comments about differences in accuracy and reliability in different makes, although none of these articles identify brands. Another issue that has turned up in the advocacy-group tests is that significant numbers of drivers have difficulty providing the required 1.5-liter samples. The breathe sample size should be reduced to 1.0 liter, because the resulting difference in accuracy is really of no consequence in a non-evidential application and the increased error associated with smaller samples will be permissive. (However, the interlocks being tested in Sweden have a very low trigger threshold). Access to results from these tests could be useful in guiding both research and implementation plans in the United States.

5. TECHNOLOGIES UNDER DEVELOPMENT

5.1 TISSUE SPECTROSCOPY

5.1.1 Description

A spectroscopes is a device that measure what proportion of a beam of light is absorbed or reflected by a sample at various wavelengths. A portion of the spectrum known as the "near infrared" (NIR) is particularly useful for quantitative analysis of organic compounds because such molecules have numerous resonances in that range. Absorption occurs at the wavelengths of the resonances. This region is called the "near infrared" because it lies just beyond the range of human vision. Most of the instruments in police labs used for evidential analysis of breath samples are NIR-absorption spectrometers.

The concentration of ethanol in tissue changes its absorption of NIR light at certain wavelengths. This phenomenon allows estimation of BAC by measuring how much light has been absorbed at particular wavelengths from a beam of NIR reflected from the tissue of the subject. Infrared light easily penetrates several millimeters of tissue so the reflected signal reveals information about the tissue to that depth. This makes NIR reflectance spectroscopy relatively insensitive to contaminants on the surface of the skin. Because the reflected spectrum is affected by many other chemicals present in the skin, the estimation relies on a complex statistical process called a partial-least-squares model. A regression analysis of the reflectance spectrum from the subject's skin is performed against a matrix of a few hundred spectra from samples with known BACs. The invention of this technology is described in two articles in the *Journal of Applied Spectroscopy* published in 2005 by Trent Ridder *et al* of InLight Solutions, Inc.[52, 53] The physics and engineering of tissue spectroscopes are complex; a convenient summary is provided in a paper by Simon Ghionea.[54]

InLight Solutions has licensed its patents and technology to its subsidiaries, Lumidigm, Inc., and TruTouch Technologies, Inc. Lumidigm is developing low-cost biometric sensors for personal identification and may be able to derive useful measures of alcohol concentration from them, although that is not the primary objective of its current development efforts. TruTouch is pursuing high-end clinical and evidential applications.

In fall, 2007 the Bernalillo County (NM) Sheriff's Department is scheduled to begin testing a prototype roadside BAC measurement tool that uses NIR spectroscopic technology.

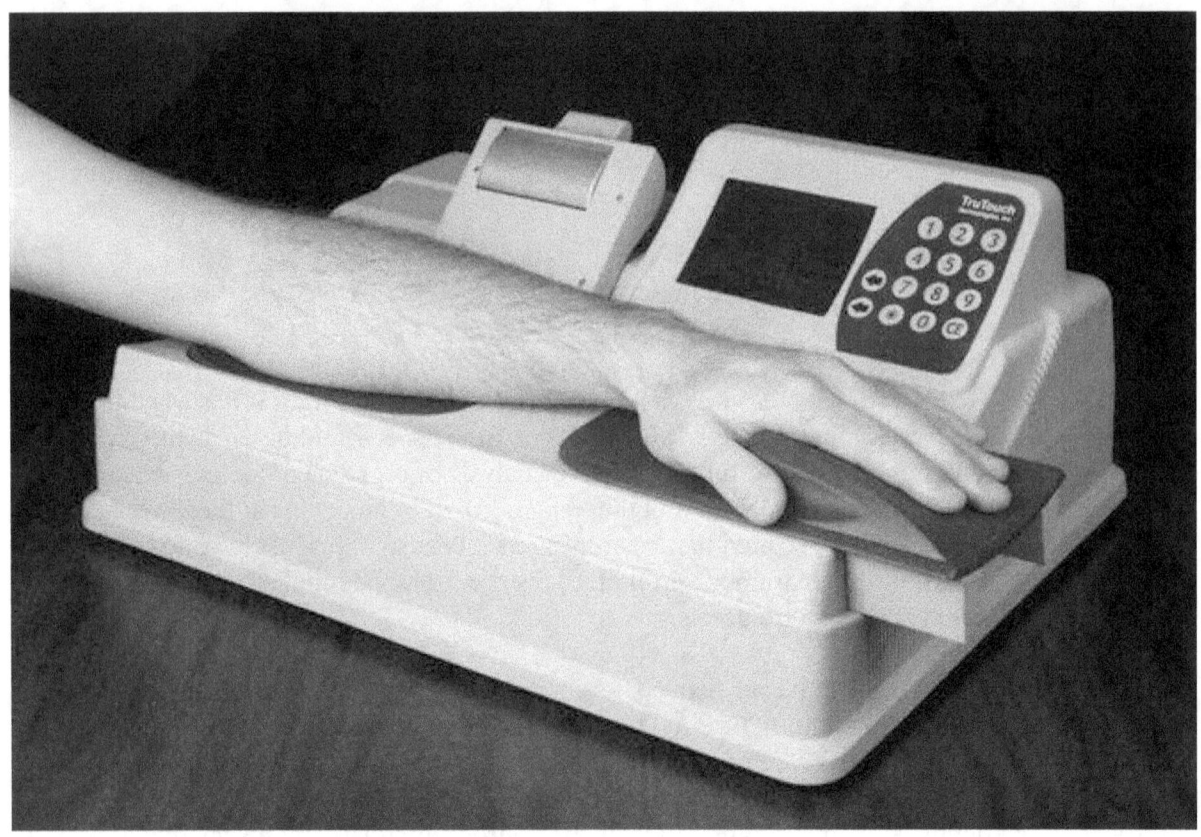

Prototype of the TruTouch Evidential NIR spectrometer

5.1.2 Validation Experiments

Initial published data[53] comparing estimates of BAC made with tissue spectroscopy against true BAC show excellent correlation, as shown in Figure 5-1. This data was based on measurements at 28 different wavelengths over a period of 30 seconds. The root mean square (RMS) error of prediction in this data is 4.9 mg/dL (80mg/dL is the per se limit on this scale). In comparison, BrAC analyzers result in RMS error of 15mg/dL (Figure 5-1),

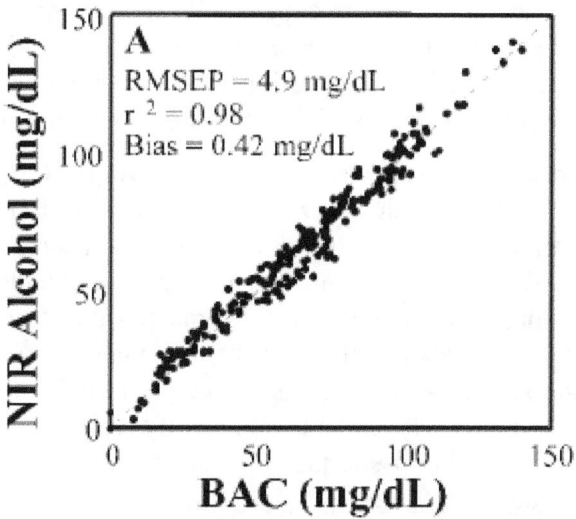

Figure 5-1 Comparison of estimates of BAC from NIR spectroscopy with true BAC

These results represent levels of accuracy, sensitivity, and specificity to ethanol that are far superior to other known methods of measuring alcohol impairment that do not involve extraction of bodily fluids.

5.1.3 Limitations

The accuracy of a statistical estimation process depends on the quantity and quality of the input data. The quantity of input is a function of the number of different wavelengths that are measured and the number of times each is sampled. The quality of the data is affected by various physical properties of the detector, such as bandwidth, noise, linearity, stability, etc. Achieving narrow bandwidths needed to avoid false positive results at low cost is particularly challenging. Reducing the size, cost, and measurement time of the tissue-spectrometer, while maintaining data quality, will require a substantial effort in technology development, testing, and refinement. Various technology developers whom we interviewed have suggested this effort will take 5 to 20 years.

For the longer term, before development of unobtrusive interlocks can begin, some physiological issues must be resolved. The soft, thin skin on the underside of the forearm works well for reflectance spectroscopy. Little is known about the reflectance characteristics of the thicker, tougher skin of the palms and fingers, or blood perfusion rates in various parts of the hand, or in the effects of the bony structures that lie close to the skin. Individual variations caused by manual labor are likely to be much larger than for the forearm.

5.1.4 Alternative Implementations

The current TruTouch prototype uses a Michelson interferometer (http://www.newport.com/-Introduction-to-FT-IR-Spectroscopy/405840/1033/catalog.aspx) as the tunable element in the system – resulting in considerable size and expense. While this configuration (shoebox size) is quite acceptable in a clinical instrument or in an evidentiary test device, it is too large and too expensive for a mass-market interlock application. The current engineering challenge is to invent devices that are much smaller, cheaper, and faster. Among the approaches suggested are:

- A large-aperture Raman spectrometer being developed by Scott McCain at the Fitzpatrick Institute of Photonics at Duke University with funding from the NIAAA. (Most spectrometers measure the amount of light absorbed or reflected at various wavelengths; Raman spectroscopy involves measurements of photons scattered at wavelengths that are different from the monochromatic incident source. These photons arise from complex quantum mechanical interactions between light and matter, and their numbers and energy levels can be used to identify molecules and determine their concentrations.) The Duke team recently reported its first success in measuring ethanol concentration in tissue-phantom samples.
  http://www.pratt.duke.edu/news/?id=235 [55]
  http://www.disp.duke.edu/publications/CSTEMPLATE_dukethesis.pdf. [56]

Prototype of the Duke large-aperture Raman spectrometer

- A Fabry-Perot interferometer (http://en.wikipedia.org/wiki/Fabry-Perot) implemented as a micro-electromechanical system, by Professor Chris Backhouse at the University of Alberta.[57]

- A distributed-feedback laser diode as the tunable element, by Simon Ghionea at Oregon State University.[54]
  http://www.nasatechnology.org/technologies/assets/2139_GSC-13915.pdf
  and http://ostc.physics.uiowa.edu/~olesberg/papers/Olesberg-2005-ProcSPIE.pdf.
- Use of discreet LEDs and filters for each wavelength, by Lumidigm.[58] (The Lumidigm work is proprietary. The following paper describes the concept: http://nr.stpi.org.tw/ejournal/ChiChemSociety/2006/EJ52-2006-1067.pdf.)
- Quantum-dot LEDs (QDOTs) – a new type of light emitting diode (LED) with the narrow bandwidth of a laser LED, but a much smaller physical volume. This technology is only at the conceptual stage; see:
  http://faculty.uml.edu/jtherrien/Classes/QE/Files/QD_Spectrometer.ppt#266,1 2,An Introduction to Quantum Dot Spectrometer.

At the current state of the art, all of these spectrometric techniques take too much time to measure alcohol concentration in tissue – minutes – to be viable for use in interlocks. However, because technological progress has been so rapid in the detectors and signal processors used in these devices, it is not unreasonable to expect that they may become viable in less than a decade. An improvement of at least a factor of 10 is required.

5.2 NEW TECHNOLOGIES TO DETECT ALCOHOL VAPOR

Deterrence of alcohol-impaired driving is a recognized need worldwide. Various technologies are being developed to serve this end, several of which focus on ethanol vapor detection. These are described below followed by a discussion of their applications.

5.2.1 Descriptions

A group at the Prokhorov General Physics Institute in Moscow has constructed a tunable-diode laser spectroscope (TDLS) that can detect ethanol vapor in moving vehicles that pass through its beam.[59]

The alcohol vapor detector works for vehicle speeds up to about 10 mph. With funding from the U.S. Department of Energy through Brookhaven National Laboratory, Russian research on TDLS technology is continuing, but its focus has been switched to the detection of explosives used in car bombs.

Alcohol vapors affect the optical and acoustical properties of various nanostructures, giving rise to possibilities for the development of sensitive detectors. Three Italian groups have demonstrated transducers that integrate single-walled carbon nanotubes with quartz-crystal microbalances or silica optical fibers.[60] Each of these detectors is reported to be highly sensitive and effective at room temperature. These are examples of technologies emerging from basic research which may someday be applicable to the detection of alcohol vapor.

At the Institute for Materials Science in Hanoi, researchers are investigating nano-crystalline perovskite oxides doped with strontium. Their goal is to produce very small, low-cost, solid-state detectors with specificity and other performance characteristics superior to the Taguchi cell. Their prototypes are reported to have excellent sensitivity and linearity, as shown in Figure 5-2. [61][11]

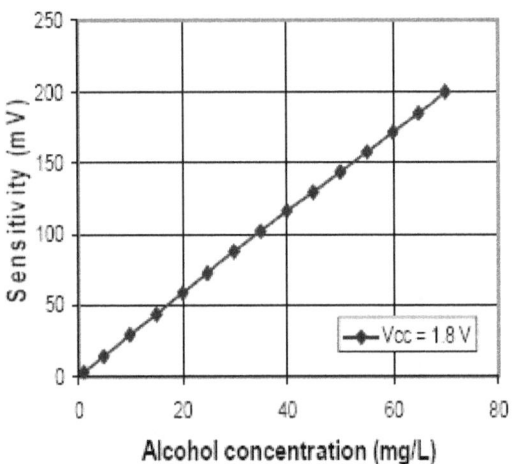

Figure 5-2 Linearity and Sensitivity Characteristics of Perovskite Oxide Sensor

In the United States Giner, Inc., has developed a variety of sensors that can continuously monitor low concentrations of alcohol vapor. The solid-polymer-electrolyte sensor (details in U.S. Patent 5,944,661 [62] may be applied as a monitor of ethanol concentration in perspiration (which is correlated with BAC). This technology or other devices Giner has built using fuel cells can also monitor alcohol presence in a vehicle. These applications are described separately in the following sections.

5.2.2 Applications

Continuous, non-invasive monitoring of BAC is a long-established need of clinicians treating alcoholics and of researchers monitoring subjects in experiments. Because a small portion of the alcohol consumed is excreted in perspiration, and BAC is correlated with the alcohol concentration in sweat, an obvious approach is to measure that concentration. Two firms have produced wireless, body-worn devices to perform that function and store the data with time stamps along with a means of downloading the data to a personal computer.

The older device, called SCRAM (Secure Continuous Alcohol Monitor), is made by Alcohol Monitoring Systems, Inc., (AMS) and uses a fuel cell sensor in an ankle-worn housing. It has been used successfully on thousands of individuals to monitor compliance with court-ordered "Do not drink" mandates. It includes anti-circumvention features and a built-in, wireless inter-

---

[11] The ordinate is mislabeled in this scan of the original document. The ordinate shows detector output, not sensitivity.

face to a modem that relays the data periodically to AMS, which distributes reports to the agency concerned with the patient.

The more recent device, developed by Giner, is currently in the prototype stage. Small enough to be wrist-worn, it is appropriately dubbed WrisTAS, for Wrist Transdermal Alcohol Sensor. Current prototypes record alcohol concentrations at programmed intervals and store data until downloaded via cable. A wireless version is currently in development under funding from NIAAA.

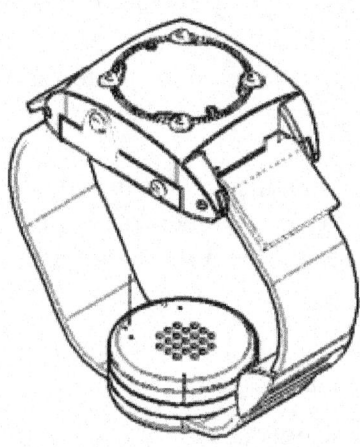

SCRAM monitors

Griner, Inc. WrisTAS sensor drawing

5.2.3 Limitations

According to interviews conducted with technology developers for this project, neither of these devices can provide an indication of true BAC that is as accurate as that from a fuel-cell BrAC monitor. Because alcohol takes from 30 minutes to two hours to appear in perspiration, the readings on the transdermal monitors lag behind true BAC. They underestimate true BAC while it is rising and overestimate true BAC when it is falling. They are also subject to reading errors caused by alcohol-containing skin-care products, but these patterns in the data stream can easily be recognized as such.

Despite these limitations, we believe that the ability of these devices to provide continuous monitoring with automatic, wireless reporting of data is so valuable in clinical studies and treatment programs for alcoholics that they are likely to find significant markets. However, their average daily costs are substantially higher than those of competing technologies for primary interlocks, such as tissue spectroscopy or fuel cells. Because they are body worn, there is a need to recharge or replace the battery fairly frequently. The Giner device must be removed for showering or swimming; the SCRAM tolerates water, but our interviewees have said that its accuracy degrades when wet. Other possible shortcomings are being explored in on-going research being conducted by PIRE. [63]

5.3 ENVIRONMENTAL MONITORS

Roadside checkpoints are an effective means of deterring drunk driving. However, since the proportion of drivers who have been drinking is usually very small, this approach consumes substantial police resources and wastes time for non-drinking motorists. Police departments worldwide are seeking ways to identify vehicles that have a significant probability of containing a drunk driver so that they can focus their attention on such vehicles and wave the rest through the checkpoint.

One concept employs alcohol-vapor sensors installed in vehicles that can communicate their data to police. The data stream would contain vehicle identifiers as well as alcohol concentrations. A low-cost, short-range service such as WiFi Max or similar would be used as the link. Police could use notebook computers or personal digital assistants (PDAs) to receive the data.

Giner, Inc., conducted tests in stationary vehicles in 2005 to demonstrate that ethanol concentrations from open containers of beer can be reliably detected within a few minutes with the windows half open and the air-conditioning system set to recycle.

No demonstrations have yet been conducted in moving vehicles, but numerous questions have been raised, among them:

- How does the concentration of alcohol vapor in a moving vehicle vary with window and/or ventilation openings?
- What are the effects of vehicle speed and air temperature?
- How can the trigger point of the detection system be adjusted to compensate for these effects?
- Can the detector be "poisoned" by the deliberate introduction of other chemicals into it?
- How can circumvention by plugging the detector be prevented?
- Can the detector withstand the rigors of the automotive environment?

5.4 VEHICLE-BASED IMPAIRMENT MONITORS

5.4.1 Use of Physiological and Vehicle Sensors to Detect Alcohol Impairment

Section 4.1.3 reviewed evidence that indicates a partial dissociation between per se and behavioral definitions of alcohol impairment. Countermeasures that detect impaired driving through objective behavioral measures could decrease crash risk beyond what is possible through direct alcohol detection and a per se BAC criterion. They could represent a second line of defense against alcohol-impairment crashes.

Stapleton, Guthrie, and Linnoila[64] reviewed the evidence indicating that alcohol and other drugs affect eye movements and considered its relevance to driving. Generally, simulation studies found that alcohol lengthens dwell times, reduces the frequency of eye movements, and also reduces the frequency of eye movements toward areas at a distance from the front of the vehicle and peripheral to the direction of vehicle motion. Alcohol slows saccadic eye movements, which are involved in bringing an object to the center of the visual field, and impairs smooth-pursuit eye movements. It results in horizontal gaze nystagmus (HGN) about 30 minutes following alcohol intake. Nawrot, Nordenstrom, and Olson[65] describe HGN as follows: "Gain, the ratio of eye velocity to target velocity, should be very close to 1 if eye movements are to maintain fixation on a moving target. Intoxication reduces the velocity of slow eye movements, yielding a gain less than 1, and thereby requiring the visual system to recruit fast eye movement to generate a 'catch-up saccade.' This produces the jerky eye movements, also called horizontal gaze nystagmus" (p. 859). The visual angle at which HGN occurs corresponds approximately to BAC. [66] Stapleton et al.[64] also reviewed the effects of alcohol on vergence, which occurs when the eyes move in opposite directions to produce binocular clarity. Alcohol produces exophoria (outward movements) at viewing distances corresponding to fixations on in-vehicle displays and esophoria (inward movements) at distances corresponding to fixations on traffic or road signs.

Stapleton et al.[64] compared the effects of alcohol, cannabis, benzodiazepines, barbiturates, and methadone on their effects on saccadic eye movements (maximum velocity, latency, accuracy), smooth pursuit (maximum velocity, gain), nystagmus, and vergence. Each drug category was described as resulting in slowed maximum velocity, undershoot, decreased gain, production or modification of nystagmus, impaired vergence, or no change. The effects of alcohol and barbiturates on these eye movement categories were identical, but distinct in two or more effects from the other three drug categories.

HGN and other ocular measures are used by drug recognition experts (DREs) who are "called to examine suspects who are believed to be under the influence of drugs, but who do not have BACs sufficiently high to justify a charge of driving under the influence of alcohol."[66] The first step of the 12-step DRE procedure is to measure BAC. If alcohol intoxication is ruled out, the remaining 11 steps are performed, including two that test ocular responses. Step 4 involves eye exams that include testing for horizontal and vertical gaze nystagmus and lack of convergence. Step 7 (a dark room examination) includes measurements of pupil size. These ocular characteristics and other physiological signs are used to identify the type of drug. Horizontal gaze nystagmus, for example, is present with depressants, PCP/phencyclidine, and inhalants. Lack of ocular convergence and normal pupil size are also generally present with these three drug types. Dilated pupils are present with stimulants, hallucinogens, and (generally) cannabis. In contrast, constricted pupil size is only present with narcotic analgesics. Additional measurements taken as part of the assessment are pulse rate, blood pressure, body temperature, muscle tone balance, body sway, and nose touching. Attempts to validate DRE evaluations that were reviewed in this report showed high error rates. For example, one showed a 62 percent false alarm rate in detecting impairment. Sources of inaccuracy include human factors and the validity of the measures.

Vehicle measures of ocular and eye movement performance could reduce human error in impairment detection and automatically institute vehicle countermeasures if sufficient evidence of impairment is present. However, even with recent advances in ocular and pupil capture technologies that could support in-vehicle implementation, "glance based measures are notoriously difficult and/or time consuming to collect and analyze. This is the reason why they are not as frequently used as the vehicle performance measures."[67] Victor et al. reported that measures of gaze toward the road center (percentage and variability) were easier to obtain. The evidence reviewed by Stapleton et al.[64] suggests that many ocular and eye movement measures may also demonstrate sensitivity to alcohol impairment even though they are unlikely to distinguish alcohol impairment from impairment resulting from other sources and currently would represent a challenge to implement.

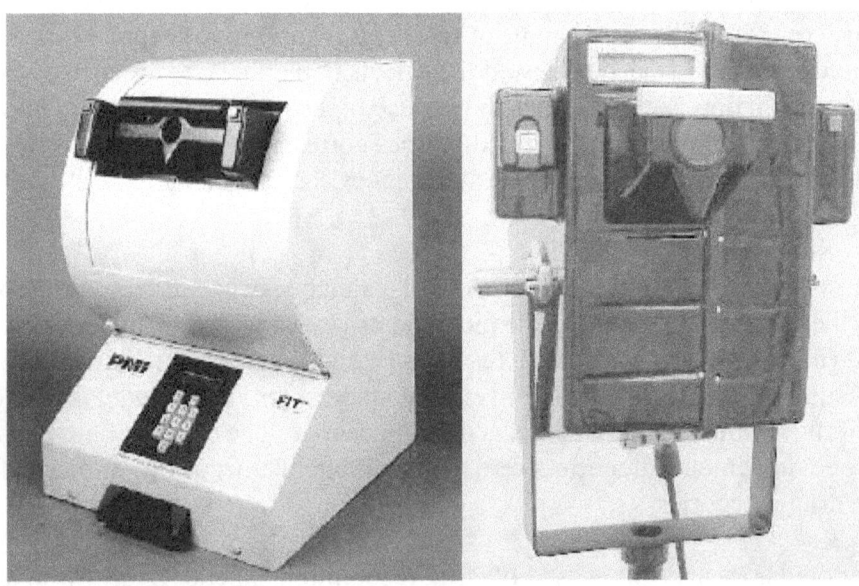

Laboratory and portable versions of instruments for pupilometry and saccadic velocity measurement

Physiological impairment detection methods that currently require electrodes attached to the driver's body (not necessarily to the head) are included in this survey because of the potential shown for tissue spectroscopy and the potential application of in-vehicle telemetry to replace electrodes. Brookhuis[68] described the driver status monitor as "non-intrusive in-vehicle measures that might be used to monitor driver state continuously...using a small set of in-vehicle sensors...to undertake proper action to forestall accidents or alert the car driver" (p. S64). A key assumption behind this concept is that the sensors will function in a non-intrusive manner. A wireless EEG manufacturer currently offers to provide "validated measures of engagement, mental workload, and distraction/drowsiness." The manufacturer describes the device as follows:

"The sensor headset can be easily applied and comfortably worn for over 8 hours of continuous use. The patented EEG sensor dispenses a small amount of conductive cream through the hair to make electrical contact, which eliminates the need for hair or scalp preparation. ... Wireless EEG allows the user freedom to move without generating artifacts obtained with conventional, wired EEG systems. The sensor headset can be easily covered by a baseball cap if worn in public."[69]

Wireless Sensor Headset: the patented wireless sensor headset provides EEG and electro-oculographic (EOG) signals

Source: www.b-alert.com/EEG.html

In 1991, De Waard and Brookhuis[36] reported a feasibility experiment for a device that would provide continuous driver status monitoring of physiological and vehicle parameters as part of the European Project DREAM (Driver Related Evaluation And Monitoring). Subjects drove in traffic with instructions to keep a constant lateral lane position and speed in some parts of the procedure and in others to follow an instrumented vehicle at a safe distance while it accelerated and decelerated. Both physiological and vehicle measures were obtained. The authors compared the onset of changes in these measures and their sensitivity to the independent variables, alcohol (BAC < .05) and time-on-task (total drive duration was 3 hours). The physiological measures consisted of various heart rate measures and an EEG activation index (theta + alpha/beta). Alcohol resulted in effects on heart rate measures, time-on-task did not. EEG activation in the alcohol condition showed a non-significant increase. Lane position was sensitive to alcohol while steering wheel variability was not. The authors determined that "these physiological signs of changes in driver status are readily followed by changes in driver behavior" (p. 304), citing a figure showing that EEG activation reached a plateau prior to an increase in lane position variability. EEG activation did not reach a plateau in the alcohol condition suggesting that a plateau would occur later. The authors describe the effect of alcohol on EEG during the driving tasks as: "It does not cause a gradual change but starts at a different offset" (p.305).

Recent physiological breakthroughs may help to achieve Brookhuis' concept of negligibly intrusive impairment monitoring.[68] To date, research on the following physiological events has only examined sensitivity to task load manipulations. Functional near infrared monitoring of brain hemodynamics has been explored as a means of monitoring cognitive and emotional states during demanding tasks. [70] The source of the fNIR data in this study was a flexible sensor that was applied to the entire forehead. Blood oxygenation closely tracked some task load manipulations.

In Izzetoglu et al.[70], 20 psychophysiological measures were evaluated as potential "gauges" of task load in a simulated military command and control exercise. St. John, Kobus, Morrison, and Schmorrow[71] discussed their effectiveness. The results indicated that both fNIR and the continuous EEG measures were sensitive to a task load manipulation. Other measures that demonstrated sensitivity to task load were directly associated with task performance, such as mouse clicks, or were evoked cortical responses to significant task events, such as error feedback sounds. Galvanic skin response (GSR) did not gauge task load in this study.

The ultimate objective of these studies, which are part of the Augmented Cognition Program of the Defense Advanced Research Projects Agency, is to integrate the successful gauges into a metric that is accurate enough to "automatically execute workload reduction systems."[72] Alcohol impairment causes timesharing (i.e., multiple task performance that requires divided attention), to become more difficult[73], suggesting that these measures may be especially sensitive to the workload of alcohol-impaired drivers.

To summarize, the studies reviewed in this section demonstrate progress in the evolution of physiological measures that have potential for in-vehicle use. A key requirement is that they operate essentially unobtrusively. Ocular, gaze, and eye movement measures have demonstrated sensitivity to alcohol impairment, but their implementation in vehicles is a challenging problem. Other physiological measures such as EEG have been reported to predict impaired driving due to a moderate dose of alcohol. Progress toward reducing their obtrusiveness has occurred in recent years, although sensors must still remain in contact with the head. Changes in blood oxygenation detected by fNIR technology have been shown to correspond to changes in task load, but their sensitivity to alcohol impairment has not been reported.

5.4.2 Use of Physiological and Vehicle Sensors to Distinguish Alcohol from Other Impairment Sources

Perhaps the most diagnostic physiological indication of alcohol impairment is HGN, currently used to screen drivers for alcohol impairment and other drug intoxication. However, integration of HGN measurement into vehicles would pose a difficult technical challenge. The review by Stapleton et al.[64] indicated that no individual or group of ocular measures can distinguish impairment due to alcohol from impairment due to barbiturates. Similarly, lane position variability is sensitive to the effects of numerous substances, which permits its use as a "standard road test" for pharmaceutical drug evaluations to determine whether their effects pose a hazard for drivers.[74, 75] These studies employ a "benchmark" strategy[76, 77] that gauges the extent of impairment associated with a psychological state by comparing it to performance from BAC at a per se limit. Such comparisons have been conducted for alcohol and particular benzodiazepines [78, 79]; dextroamphetamine[79]; cannabis[34, 80], visual impairment[44, 81], fatigue[79, 82, 83]; and mobile or cell phone use while driving.[76, 77, 84]

Some of these studies have found results that discriminate alcohol from other impairment sources on laboratory task performance. For example, Tiplady et al.[78] found distinct profiles for the effects of alcohol and temazepam. However, efforts to convert laboratory tasks for use in a roadside testing device that would distinguish alcohol from other states have met with limited success thus far: Tiplady, Dewgia, and Dixon[85] report that only 60 percent of individuals in an alcohol condition (BAC = .08) exceeded the 75th percentile (in the direction of impairment) of their impairment scores in the placebo condition.

Strayer et al.[77] found that vehicle parameters distinguished between alcohol (BAC = .08) and cell phone use in a simulated car following task. Slower braking responses to taillight onset in the lead vehicle and more collisions due to failures to brake were found in the cell phone conditions, whereas alcohol produced closer following and harder braking. These results for the alcohol conditions may be related to findings that Terhune et al.[33] reported. Terhune et al. described conditions (speeding, lane maintenance, inattention, right-of-way) under which fatal crashes related to alcohol and other drugs occurred. Their results suggest that speeding may represent part of a unique "signature" for alcohol impairment. This conclusion would apply to the effects of high BAC levels characteristic of those found in fatal crashes involving alcohol.

In his TNO interview, Ramaekers said that he has found evidence, in his own research as well as in research for the EU Project IMMORTAL that driving behavior under the influence of alcohol is characterized by an increase in lane position variability, a shift in the vehicle's lateral position toward the center of the road, increases in average driving speed and speed variability indicative of greater risk taking, and a delay in motor actions and responses, such as braking reaction times. Ramaekers believes that sedative drugs lack the risk-inducing potential of alcohol. It could be hypothesized – although this hasn't actually been tested yet – that the increase in speed and speed variability would be absent for these drugs, while the other effects would remain. When it comes to stimulants like ecstasy the pattern would, again, be different. However, the strongest effects would now probably be found in cognitive impairment, in particular in verbal and spatial memory. In order to assess this one would have to add elements covering these tasks in the experimental assessment of effects.

Infrared spectrometry is widely used in vitro for identification and quantification of an enormous variety of chemicals -- including substances of abuse and metabolites associated with fatigue. However, such analyses are performed on blood samples, usually with various preparatory steps in which the sample is refined and stripped of extraneous components before the spectrometry is performed. With these prepared samples, microscopic quantities can be identified and measured.

In contrast, tissue spectrometry is performed in vivo. Without the advantages of sample extraction and preparation, the noise floor of the spectrometry is raised by several orders of magnitude. Only analytes that are present in large quantities, such as glucose, can be measured by this technique. Because alcohol must be consumed in substantial quantities to impair driving — usually more than 100 grams — it can also be measured by tissue spectroscopy. This is not the case for other substances of abuse or for biomarkers associated with fatigue where the quantities present in the whole body are generally reckoned in milligrams.

It may not be worthwhile to uniquely identify alcohol impairment. In his interview with Ward, Brookhuis expressed the view that identifying the source of impairment could be counterproductive because "increasing the categories of impairment that must be separated makes it harder to get a 'hit,' so the accuracy of a system to detect general impairment will necessarily be higher than one that has to hit separate types of impairment."

5.4.3 Vehicle-Based Impairment Detection Using Multiple Sensors

Most of the studies discussed thus far provide evidence on the power of individual metrics to identify impairment. One effort, Project SAVE (System for effective Assessment of the driver state and Vehicle control in Emergency situations), made innovative use of the combined output of multiple driver and vehicle sensors as inputs to an integrated monitoring unit (IMU). SAVE recorded eye blink, eyelid closure, and steering wheel grip data using vehicle-based driver sensors. It also recorded sensor outputs regarding the vehicle. These data included lateral position, lateral position variability, time to lane crossing, and steering wheel angle and variability. The output of the IMU was the driver's putative state: impaired or unimpaired. The effectiveness of the algorithms generated by SAVE was validated against several sources of impairment including alcohol and fatigue. According to Peters and Van Winsum[3] the alcohol and fatigue evaluation aimed to "evaluate whether the system is able to distinguish between drowsiness-induced impairment and alcohol-induced impairment" (p.45).

SAVE issues warnings if impairment is detected. If there is no reaction to a warning or if the situation is considered critical, SAVE "will stop the car safely alongside the road. This requires no human involvement and is carried out on basis of the information provided by sensors indicating the presence of other vehicles and relevant objects and the car's current lane position" ([3], p. 17). This countermeasure was only considered acceptable when respondents understood that its use was limited to situations where the driver failed to respond.

To define impairment for the purpose of detection (an operational definition), it would normally be necessary to set threshold values for vehicle-based sensors that can successfully distinguish between impaired and unimpaired driving. For example, Brookhuis, De Waard, and Fairclough [86] proposed thresholds for vehicle sensor metrics. SAVE avoided the use of explicit thresholds by having driving instructors accompany the subjects and provide expert ratings of impairment every five minute epoch during the closed course drives used to help "train" the neural network IMU.[3] Two driving instructors provided quantitative ratings of impairment during a fatigue inducing four-hour drive, the first part of which was considered normative or baseline driving, and one instructor rated impairment during a one-hour alcohol-dosed drive. Sensor thresholds were not preset and were not identifiable because of the type of IMU employed. Alcohol measurement was not used to train the neural net. The neural net processed the driver and vehicle data in conjunction with the instructor ratings during the neural net training drives to develop the neural net's impairment detection capabilities.

Three IMU training regimes were studied in a closed course evaluation:

(1) A personalized or individualized situation-specific regime was based on individual driving performance under baseline, alcohol, and fatigue conditions, and obtained separate algorithms for each impairment condition for each subject. It was used to classify the driving of the same individuals.

(2) A generalized regime trained the neural net using the results of three subjects and then used the results to classify the remaining six subjects' driving.

(3) A personalized non-situation-specific regime trained the neural net only under baseline, or normative conditions, and later applied it to the same individuals.

The goal of Project SAVE was to correctly detect impairment in at least 90 percent of the 30-second epochs while maintaining a false positive rate of one percent or less.

In his summary of the SAVE results on the detection of alcohol impairment, Ward[87] adapted results from Peters and Winsum[3] to arrive at average performance values for the three training regimes that were compared (see Table 5-1). Percentages refer to detection in a series of five-minute epochs. The two personalized training regimes produced comparable results. Using both normal and impaired driving in the personalized training regimes only benefited sensitivity by about 5 percent, compared to using only normal driving. The generalized regime resulted in less sensitivity. According Ward's description of the results, the personalized SAVE regimes performed better when identifying impairment due to fatigue than alcohol. For example, the condition-specific regime correctly classified 84.5 percent of the fatigue epochs, with 4.6 percent false alarms and 10.9 percent misses. However, the generalized regime did little better when classifying fatigue: it demonstrated 45.2 percent accuracy with 21.3 percent false alarms and 33.5 percent misses.

Table 5-1 Average Performance of Project SAVE Classification Regimes

Classification Regime	Accurate	False Alarm	Miss
Personalized Condition –Specific Regime	64.6%	12.5%	22.9%
Personalized Normative Regime	58.6%	14.3%	27.1%
Generalized Regime	40.7%	27.4%	31.9%

In the Project SAVE final report[88] Bekiaris stated that the overall accuracy for detecting impairment was as follows (with false alarm rates in parentheses): 95 percent (1.2 percent) accurate for detecting a sudden loss of control/illness, 93 percent (2.1 percent) accurate for detecting impairment due to inattention, 88 percent (4.6 percent) accurate for fatigue, and 78 percent (8 percent) accurate for alcohol (BAC = .05).[12] Bekiaris attributed the alcohol detection results to the use in the evaluation of the applicable per se limit (BAC = .05), "which is a very mild rate, hence they do not represent serious impairment and thus the relevant detection is rather low" (p. 13). These results were obtained with personalized impairment definitions. This summary of the results is all that is publicly available because the report documenting the Project SAVE results (Deliverable 10.3.2) remains confidential. Thus the length of time the IMU took to detect impairment was not reported, nor was the ability of the IMU to discriminate alcohol from fatigue impairment (the reported results only concerned the ability of the IMU to distinguish alcohol from placebo conditions).

While the results of Project SAVE emphasize the sensitivity advantages of personalized definitions of impairment over general definitions, research is needed to establish effects on

---

[12] Most European nations set the per se limit equal to BAC = .05.

bottom line measures such as crash risk. The issue is that a personalized algorithm would most likely classify as unimpaired many drivers who exhibit greater crash risk than drivers it classifies as impaired. A general definition of impairment based on crash risk avoids these potential misclassification errors. The following considerations favor general definitions of impairment:

- The personal definition would not classify driving according to crash risk. The same driving sample could be classified as impaired or unimpaired. It could be classified as impaired if exhibited by a good driver, and as unimpaired if exhibited by a poor driver.

- The general definition could classify vehicle behavior strictly according to crash risk or according to an alcohol "signature" that is valid for all drivers. Driving that exhibits the alcohol signature or that is risky would be classified as "impaired" even if it does not differ from a (poor) driver's typical driving. A skilled driver would be classified as unimpaired when his or her crash risk is sufficiently low, even when it is worse than usual (e.g., due to road or weather conditions).

- Since it would correspond directly to crash risk, the general definition would agree with the common sense definition of impairment as "bad driving." This would increase its acceptability.

- The general definition would be applicable to all vehicles regardless of the driver whereas the personal definition would be unique to each driver. Thus, the general definition could be hardwired into vehicles, whereas the personal definition would require a means to transfer the individual's impairment "signature" to a different car. Otherwise, it would be necessary to develop a new signature each time the driver uses a different car. This would be an important criterion for rental car fleets where the vehicle would have relatively little opportunity to acquire a personalized alcohol impairment signature.

In the TNO interview, Ramaekers "stressed the importance of individualization of baselines when assessing the effects of alcohol, drugs, and medicines on driving behavior. The 'natural' behavior of the driver should be known before additional effects can be estimated." This is true so long as one defines impairment relative to the individual, but greater benefit could occur if one combines the personal definition that uses an individual baseline with a general definition that defines impairment relative to the general population. The main reason for expecting additional benefit from a general definition is that it would not miss impaired driving that does not differ appreciably from a poor driver's unimpaired driving, but which nonetheless exhibits unacceptable risk.[13] A relative definition of impairment may be appropriate for understanding the effect of alcohol or a new medication on driving, but may not be appropriate for defining impairment for the purpose of reducing impairment crashes. Accordingly, a combination of non-situation specific individualized and generalized baselines should represent the ultimate objective for behavioral TOPICs. When the individualized alcohol signature is encountered, the generalized crash risk baseline thresholds could be reduced.

---

[13] Harrison and Fillmore found a moderate negative correlation between unimpaired driving skill and the size of the impairment decrement, suggesting that there may be relatively few cases where BAC = .08 (the alcohol level they studied) does not produce a detectable decrements among less skilled drivers.[16]

Regarding the current status of Project SAVE and its successors, TNO interviews summarized recent developments and plans for the EU vehicle sensor research program, which included SAVE. The more recent research (AWAKE, SENSATION[14]) did not evaluate vehicle detection of alcohol impairment. However, Vits of the EU DG INFSO expects SENSATION to yield the final solutions to on-line driver monitoring, possibly leading to an FOT and a stage in which the findings will be implemented and [the technology] miniaturized before they will go into production.

Project SAVE provides an example of vehicle-based impairment detection using multiple behavioral sensors. Expert judgment was used to train a neural net to identify impairment instead of using explicit threshold values that would distinguish sensor inputs of impaired drivers from those unimpaired drivers. Three regimes were used to train the neural net and in effect establish a baseline against which to compare later driving samples: two personalized regimes and one generalized regime were used. The personalized baselines performed better than the generalized baseline, but they would classify driving behavior in a way that does not necessarily correspond to crash risk. For example, they could classify the same driving behavior as impaired or unimpaired depending upon what is typical for that driver.

The primary benefit of behavioral definitions compared to "primary" ignition interlock technology is that they might detect impairment caused by BAC levels less than the per se limit, such as impairment resulting from low levels of alcohol combined with fatigue or other factors. However, some amount of driving is required before detection occurs so unlike ignition interlocks, they cannot prevent impaired driving.

---

[14] AWAKE is concerned with detecting and warning drivers of fatigue impairment independent of driving conditions. SENSATION is intended to use the results of AWAKE to build a reliable and robust fatigue-monitoring system. [89]

6. COMPARISON OF TECHNOLOGIES IN USE AND UNDER DEVELOPMENT

In this section, all of the technologies described in the two previous chapters are ranked according to technical criteria. All of the estimates are subjective and were made by Volpe staff, largely on the basis of personal conversations with the technology developers.

- *Accuracy* – precision in estimating true BAC in a vehicle when and if the technology is developed to its physical limits.
- *Cost* – unit cost for fully developed technology in mass production.
- *Development time* – years to reach mass production of units with characteristics acceptable to general public.
- *Convenience* – ability to perform BAC estimation with very little effort, distraction, or wasted time.
- *Circumvention risk* – relative vulnerability of sensor to being fooled into providing a low estimate of BAC.
- *Technical risk* – risk that the technology will never reach the mass-market, either because of unforeseen technical limits or because costs cannot be reduced sufficiently.

The following table shows a rank ordering comparison of all the technologies in relation to these technical criteria.

Table 6-1 Comparison matrix for primary interlock applications

Technologies	Criteria					
	Accuracy	Cost	Development Time	Convenience	Circumvention Risk	Technical Risk
Tissue Spectroscopy	+++	?	-	+++	++	--
BAIID	++	+	+++	-	+++	++
Transdermal	+	-	+	-	+++	+
Environmental Vapor	--	++	+	+++	---	+++
Behavioral (Ocular)	+	--	++	++	-	++
Vehicle-Based	-	++	-	+++	--	---

Scale: Best +++ to Worst ---

6.1 DESCRIPTION OF TECHNOLOGIES AND RANKING BY ACCURACY

Accuracy is a fundamental requirement in any system to prevent alcohol-impaired driving. While a direct measure of performance impairment might seem ideal, such measures are inherently technically complex, expensive, and time consuming. The "gold standard" of alcohol impairment has become BAC. It is an absolute standard, i.e., no compensation factors or comparisons against an individuals' baseline are necessary. All other approaches are tested for accuracy against laboratory BAC measurements. The rank ordering of other known methods of estimating alcohol impairment is as follows:

Tissue Spectroscopy

Tissue spectroscopy measures the concentration of ethanol in tissue by its absorption of NIR light at certain wavelengths. This phenomenon allows estimation of BAC by measuring how much light has been absorbed at particular wavelengths from a beam of NIR reflected from the skin of the subject. Because the reflected spectrum is affected by many other chemicals present in the skin, the estimation relies on a complex statistical process called a partial-least-squares model. A regression analysis of the reflectance spectrum from the subject's skin is performed against a matrix of a few hundred spectra from samples with known BACs.

The accuracy of a statistical estimation process depends on the quantity and quality of the input data. The quantity of input is a function of the number of different wavelengths that are measured and the number of times each is sampled. The quality of the data is affected by various physical properties of the detector, such as bandwidth, noise, linearity, stability, etc.

Breath Alcohol Concentration (BrAC) Ignition Interlocks

The current specification for accuracy in breath alcohol ignition interlock devices is that they must lock out at least 90 percent of the time (18 out of 20 trials) when the alcohol concentration in the test sample exceeds the set point (normally .025) by .01. At extreme temperatures or under other conditions of stress, the allowable deviation from set point increases to .02 BAC.

Most interlock manufacturers currently use **fuel-cell sensors**. This technology is fairly rugged and ethanol specific, but the fuel cell must be warmed up to breath temperature to meet the accuracy specification. This necessitates a heater assembly and significant energy use for heating — not a problem for a device that is hardwired to a vehicle, but a major barrier to the use of fuel cells in wireless devices like key fobs.

For applications requiring small size and low battery drain, **solid-state sensors** are used to measure breath alcohol. When freshly calibrated, they can be almost as accurate as fuel cells, but they show considerable drift over time. Furthermore, they respond to several volatile organic compounds other than ethanol. Since their use for enforcement purposes is not sanctioned, there is little public-domain data regarding their accuracy. Contamination is also a problem. According to the technology developers interviewed, recent research suggests substantially improved accuracy and specificity may be obtained by replacing the tin-oxide sensor with one constructed from perovskite crystals doped with strontium, but no complete monitors with this technology are yet available for testing.

Transdermal Alcohol Monitoring

Transdermal non-invasive monitoring of BAC is a long-established need of clinicians treating alcoholics and of researchers monitoring subjects in experiments. Because a small portion of the alcohol consumed is excreted in perspiration, and BAC is correlated with the alcohol concentration in sweat, an obvious approach is to measure that concentration. Two firms have produced wireless, body-worn devices to perform that function and store the data with time stamps along with a means of downloading the data to a personal computer.

The older device, called SCRAM is made by AMS and uses a **fuel cell sensor** in an ankle-worn housing. It has been used successfully to monitor thousands of individuals to monitor compliance with "Do not drink" mandates. It includes anti-circumvention features and a built-in, wireless interface to a modem that relays the data periodically to AMS, which distributes reports to the agency concerned with the patient.

The more recent device being developed by Giner, Inc., uses a **solid-polymer-electrolyte sensor** and is currently in the prototype stage. Small enough to be wrist-worn, it is appropriately called WrisTAS.

Neither of these devices can provide an indication of true BAC that is as accurate as that from a fuel cell BrAC monitor. Because alcohol takes from 30 minutes to two hours to appear in perspiration, the readings on the BAC monitors lag behind true BAC. They underestimate true BAC while it is rising and overestimate true BAC when it is falling. They are also subject to reading errors caused by alcohol-containing skin-care products, but these patterns in the data stream can easily be recognized.

The readings obtained by these devices are affected by individual differences in sweating rate, skin thickness and permeability and should be calibrated to each individual user.

Vehicle-Based Impairment Monitors

Vehicle-based impairment detection involves a multiplicity of sensors feeding a neural network or some other impairment detection algorithm. In the SAVE experiments, the variables included:

- Eye blink,
- Eyelid closure,
- Steering wheel grip,
- Mean lane position (relative to right lane marking),
- SD of lane position,
- SD of steering wheel position,
- Mean speed,
- SD of speed, and
- Time to lane crossing.

This data was processed through a neural net designed to render a decision as to whether the driver was impaired. The project leaders had hoped to obtain false-alarm rates on the order of 1 percent of the 30-second sampling epochs, non-impaired driving would be scored as impaired. The target for correct detections of impaired driving (at ~.05 BAC) was at least 90 percent. Unfortunately, the best result actually obtained was an 8-percent false-alarm rate, which implies about four false alarms per hour. Performance was substantially worse without calibration of the neural net for each subject.

Since this false-alarm rate is orders of magnitude higher than that of current breath-alcohol ignition interlocks, the driver-performance approach has not been considered for interlocks. However its sensitivity may permit use in warning systems.

Ocular Measures

One ocular measure, the horizontal gaze nystagmus test, has long been used to detect alcohol impairment. Thousands of police officers have been trained to perform the test, which requires nothing more than a flashlight and careful observation of the subjects' eyes for the jerky movements that characterize nystagmus. However, the accuracy of the field sobriety test ranks far below that of breath analyzers.

Instruments that can perform automatic measurements of saccadic velocity and pupil reactions to flashes of light have come to market in recent years. Because they must shield the subject's eyes from ambient light and perform several hundred measurements per second to obtain valid measurements, these instruments have been expensive and bulky. They have proven to be cost-effective as screeners for drug abuse. However, the data generated by these instruments is used only to decide whether a given individual should be given a chemical test. By themselves, ocular measures are not definitive.

6.2 Costs

To achieve minimal costs in impairment detection, a system must be fabricated at low initial cost, require little or no routine maintenance, and be completely automatic. No existing device comes close to minimizing all of these costs. The best prospect for a low-cost system is one that:

- Is as simple as possible;
- Can be produced using semiconductor-fabrication techniques;
- Has no moving parts;
- Requires no recalibration or is self-recalibrating;
- Is inherently invulnerable to contamination; and
- Can tolerate the vehicular environment (temperatures, sunlight, vibration, etc.) so that little maintenance is required.

Tissue Spectroscopy

Tissue spectroscopy is still in an early stage of development. The near-term implementations of this technology are likely to occur in the evidential market, for which low cost and small size are not essential. Nevertheless, spectroscopy remains the best of the known candidates for low-cost interlocks, because it avoids the sensor-contamination and measurement-drift problems of other approaches, and because it appears possible that spectroscopic sensors can ultimately be produced almost entirely through semiconductor-fabrication techniques. These techniques are associated with a longstanding trend: the rapid decline in costs relative to device complexity that has persisted for decades.

BrAC Analyzers

BrAC analyzers (solid state) are widely available as personal screeners for less than $100 at retail. As such they do not include security features, data storage/reporting, or ignition-interlock wiring. Even if these features were added, they would still provide very economical interlocks, were it not for their shortcomings with drift and contamination.

BrAC (Fuel-Cell) Interlocks

BrAC (fuel-cell) interlocks are the dominant current technology for ignition interlocks. They cost about $900 per year per vehicle and the cost includes routine servicing and recalibration, data downloading, and reporting, etc. These costs have been stable and vary mainly with the general cost of living in various parts of the nation.

Transdermal Monitors

Transdermal monitors are more expensive than breath analyzers despite the similarity of their sensor technologies because:

- These body-worn devices must include rechargeable battery packs or disposable batteries;
- The quantity of data collected and processed is much larger; and
- Sweat contains various salts that necessitate more frequent maintenance.

The current daily cost of the SCRAM system (a fuel-cell based device) is $10 to $12 according the manufacturer's Web site. The WrisTAS is still in development and has not yet been priced.

These products offer an economical alternative to other methods of monitoring alcoholics in certain clinical research, therapeutic and court-mandated environments, but are too expensive for the interlock market.

Vehicle-Based Impairment Monitors

Vehicle-based impairment monitors have costs that are far more difficult to estimate than those of the various chemical sensors described above. At the present state of the art, these monitors require the installation of several thousand dollars' worth of sensors and processors to each vehicle. They exist only as research prototypes and evidence that they may work well enough to justify the design of production versions has not been reported.

Elements of a driving-performance monitoring system are advancing to market on their separate merits as crash-avoidance technologies. Such devices as electronic stability control, adaptive cruise control, forward-collision crash warning, and lane-departure-warning systems are available now, mostly in luxury models and usually as part of a bundle of options costing several thousand dollars. If such devices gain an appreciable market, it is possible that the simple addition of some software to the vehicle's computer system can combine the data from all of these devices in a way that provides at least a rough measure of driver impairment. The cost of such an impairment warning system would be negligible.

Ocular Measures

Current instruments for automated ocular tests cost about $20,000. Over time, advances in electronic technology are likely to make much cheaper devices possible, if there were a large market for them. Currently there is no indication such a market will develop.

The need to shield drivers' eyes from sunlight will probably require either a bulky device or some head-worn apparatus. Either of these form factors tends to set minimum costs higher than for sensors that can be made small enough to embed in other devices such as key fobs, cell phones, steering wheels, etc.

6.3 LATENCY

Latency applies to the various means of estimating BAC in two different contexts. The first refers to the delay by which the physical phenomenon being measured lags actual BAC. This latency is associated with the time required for alcohol to diffuse from the blood into interstitial fluid, alveolar air, or tissue. The second has to do with the time required to take a sample, analyze it, and report the results. Actual blood sample analysis has zero latency in the first sense by definition, but taking a sample, sending it to a lab, and getting the results back requires hours or days, rendering this approach impractical except for clinical or evidential applications.

By contrast all of the chemical sensors under consideration in this document have the capability to take a sample, analyze it, and report a BAC estimate in a matter of minutes. Their latencies are dominated by the time required for alcohol diffusion into the medium being sampled. Since the alcohol diffusion is from blood through interstitial fluid into cells, and then to the alveoli, and finally into perspiration; the rank ordering of physical latencies is obvious. The latencies of the measurement process are determined mostly by various complications associated with some of the devices.

Tissue Spectroscopy

Tissue spectroscopy can have the least overall latency because the diffusion time from blood to tissue is about 15 minutes, measurement cycles are short (30 seconds or less), and sampling can be nearly continuous if the sensor is embedded in the steering wheel.

BrAC Monitors

BrAC monitors lag only slightly behind tissue spectroscopy. They are, however, subject to errors caused by the presence of mouth alcohol, which necessitates a wait of 15 minutes if present. Fuel cell versions can take as much as three minutes to warm to operating temperature in very cold weather.

Transdermal Monitors

Transdermal monitors show a substantial variation in latency from 30 to 120 minutes.

Vehicle-Based Impairment Monitors

Vehicle-based impairment monitors have latency that varies with driving conditions. Under favorable conditions impairment may be detected within a minute; in the absence of other traffic and with poorly defined road edges, detection may be much delayed, or never occur at all if the sensors measure headway or lane position.

Ocular Measures

Test cycles last about 30 seconds in current automatic ocular test devices. Even if latency time could be reduced by reducing the number of tests used, accuracy would be reduced.

Nystagmus occurs about 30 minutes after ingestion of a sufficient alcohol dose[70] and latency to obtain an estimate of BAC is not significant.

6.4 USABILITY

Usability refers to a collection of factors that affect user acceptance of a product, such as convenience, efficiency, ease of use, physical and emotional comfort, etc. While there are no straightforward objective measures of these desiderata, the technologies under consideration for prevention of impaired driving are so radically different in their usability characteristics that it is not difficult to rank order them.

Tissue Spectroscopy

Tissue spectroscopy, assuming that the devices can be devices can be made very small, accurate, and fast, comes close to the ideal in usability. The concept is that with the detector embedded in a key fob or the steering wheel, the driver would not need to perform any special action to be tested prior to the start of driving, nor experience any delay.

Vehicle-Based Impairment Monitors

Vehicle-based impairment monitors are similarly convenient in that no special action is required. However, this approach inherently requires that the vehicle travel some distance from the point of trip origin before the detector can determine if impairment exists. In the event that this occurs, immobilization of the vehicle at this point will most likely cause much greater inconvenience to the driver than if the vehicle had remained at its initial location. Furthermore, the only known vehicle-based monitors have such high false-alarm rates as to raise major usability questions, especially if an alarm implied vehicle immobilization. If used to trigger an alert to the driver, they would at worst produce annoyance.

BrAC Monitors

BrAC monitors entail some inconvenience in that blowing a sample and waiting for the analysis takes about 30 seconds. For fuel-cell monitors, there is an additional wait for warm-up, which can be as much as three minutes in very cold weather. It is recommended that drivers pull over to perform the rolling retests, thus incurring further delays. However, in practice, most users do them while driving, potentially distracting the driver. Some users of BrAC monitors report that they feel embarrassed by having to perform the test when there are other people in their cars.

Transdermal Monitors

Transdermal monitors avoid the inconvenience of performing the breath test before every start, but impose a new set of usability issues. Because they are worn continuously, contamination of detectors is a substantial problem and activities like swimming are prohibited. They require more frequent maintenance than BrAC monitors, and their constant physical presence is uncomfortable and embarrassing for some users.

Ocular Measures

To detect impairment, ocular measures (i.e., of saccadic velocity and pupilometry) must be compared against an individual's baseline in an unimpaired condition. Establishing a baseline requires taking the test about 20 times in an unimpaired condition. However, current automated test devices fail to establish baselines for 5 to 10 percent of subjects, usually due to excessive blinking – generally a symptom of a "dry eye" condition.

Current instruments are designed for indoor use only, because direct sunlight would introduce too much stray light into the apparatus. Furthermore, the subject's eyes must be adapted to indoor illumination levels; otherwise the pupils would be too constricted to be measured. Also, the use of these devices is limited to parked vehicles. These considerations suggest major usability problems for ocular measures.

6.5 Technical Risk

This section presents an attempt to rank order the various technologies according to the level of risk as to whether they can function successfully through the full range of normal driving conditions. This discussion deals only with technical issues, although it must be recognized that the greatest risk most technology developments face is simply that they will prove too expensive to pursue to completion.

BAC and BrAC Analyzers

BAC and BrAC analyzers are established technologies that have undergone at least 50 years of improvement since invention. They have no technical risk and only a modest need for improvement in accuracy for BrAC analysis. In ignition interlock applications, BrAC data may be supplemented with other vehicle data, e.g., speed, VMT, time of days, GPS location, etc., but none of these potential enhancements has significant technical risk because they are well established technologies.

Transdermal Fuel Cell Monitors

Transdermal fuel cell monitors have been employed in court-mandated programs since 2004. Several thousand units are said to be in use. Some problems have been identified with respect to detector contamination and underestimation of BAC at low skin temperatures. Because these devices are regarded as too expensive for the general population of drivers, the prospects for resolution of these technical problems have not been explored.

Tissue Spectroscopy

Tissue spectroscopy has several implementation alternatives. The version being developed for evidential use has only modest technical risk. The much smaller, much cheaper implementations needed for interlock applications are high risk.

Vehicle-based impairment Monitors

Vehicle-based impairment monitors have not yet been demonstrated to achieve a satisfactory level of accuracy in any experiment and must be regarded as having high technical risk.

Ocular Measures

Although ocular devices are already used in laboratories and as screening devices for substance abuse in probation offices, they have such substantial problems with respect to cost and usability that there is little chance that effort will be expended to develop them for use as interlocks.

7. CROSS-CUTTING IMPLEMENTATION ISSUES

7.1 ACTIVE VERSUS PASSIVE METHODS

To date, effective methods for determining whether a driver is impaired by alcohol have required either a blood draw or active participation of the driver in some test procedure, e.g., blowing a sample or following instructions for a field sobriety test. Without the threat of arrest or the presence of a mandated interlock, we believe that the willingness of drivers to perform such acts routinely is nil.

There was consensus among the technology developers interviewed for this project that such active methods of assessing impairment are inconsistent with achieving voluntary acceptance of a technology to prevent impaired driving. The dilemma is simply that for every known detection technology, the sensitivity of impairment determination can be improved by imposing some active requirement(s). Many will not work at all without some active participation or some constraints on otherwise normal behavior, e.g., wearing gloves may be prohibited. There is at present very little understanding of how the public would react to such requirements.

7.2 PRE-START VERSUS POST-START TESTING

Ideally, an impaired-driving-prevention technology should block an inebriated driver from starting, or better yet – warn him not to try before he heads for his car. After a driver has passed a pre-start test, it is possible that he may resume drinking. It is also possible that someone who is drunk may be driving after a different person passed the initial test. For this reason, all current DUI-offender interlocks require additional tests at random intervals during the trip, known as "rolling retests." Drivers are encouraged to pull over to take these tests, but the technology developers we interviewed stated that most do not.

Some of the alcohol-monitoring technologies have the capability to perform both pre-start and post-start tests. For a pre-start failure, the obvious course of action is to lock the ignition. What to do when a rolling retest is failed is an open question. Disabling a vehicle once it is underway raises the prospects of leaving it stalled on a highway or leaving the driver stranded in a dangerous location. Current interlocks simply report the condition to authorities, which then impose sanctions on the drinking driver at a later date. Unfortunately, this process requires an elaborate infrastructure to process and act upon the data. There is at present no indication that any unit of government at the local, State, or national levels favors the creation of such a process that would apply to all drivers (as opposed to the current process applicable only to DUI offenders).

For technologies that may detect impairment based on driving performance, the same issues arise for their potential use as an ignition interlock.

7.3 CHOICE OF THRESHOLD TO TRIGGER COUNTERMEASURES

Current BAIIDs used by DUI offenders enforce a zero-tolerance policy toward driving after drinking. Their actual set points are typically equivalent to .02 or .025 BAC in order to avoid false positives due to measurement errors. In theory, a driver might be able to have one drink and still start his car, but cannot count on that because of the uncertainty in the measurement error.

We believe that there is widespread public support for prohibition of impaired driving, but no such consensus on a zero-tolerance policy. Attempts to impose one are likely to be met with strong, organized opposition. Most drivers who drink at all will refuse to purchase vehicles that embody such policies, so manufacturers will not produce them absent legislative mandate.

A set point equivalent to the per se limit, .08 BAC, is what is expected by the interviewees who commented on the subject, and what Saab and Volvo have indicated they will offer. There may be additional argument about whether the actual set point should be equal to the nominal value (say .08) plus the standard error of the measurement (to minimize false positives) or minus the standard error to minimize missed detections).

7.4 PRIVACY ISSUES AND CIRCUMVENTION

By statute or regulation current interlocks record and report to authorities data about every attempt to start a vehicle and every rolling retest. Authorities apply sanctions for failed retests after the data are reported – not at the time of the offence, because it would be dangerous to stop vehicles en route. Circumvention by disabling or bypassing is also detected and reported. In some jurisdictions, interlocks are required to include an emergency override switch to prevent the possibility of a motorist freezing to death or drowning in a flood because of a denied start. Use of this switch is reported to authorities and must be justified to avoid sanctions. Without data recording, reporting, and the threat of penalties, the effectiveness of interlocks may be seriously compromised, especially among motivated drinkers.

On the other hand, data reporting arouses a host of privacy and civil liberties concerns. Large numbers of prospective buyers who would never drive drunk may nevertheless reject technologies that report their vehicle uses to authorities. Automobile manufacturers we interviewed think the idea would be death to sales. Personal data about BAC readings appears to fall under the Health Information Privacy Protection Act (HIPPA). We have been cautioned by NIAAA that we must be very careful in planning experiments with these technologies not to run afoul of HIPPA.

In short, impaired-driving-prevention technologies may be unmarketable with data reporting to authorities and easy to circumvent without it. This issue is both critically important and extraordinarily difficult to research.

7.5 VEHICLE TELEMATICS

The term "vehicle telematics" refers to the sending, receiving, and storing of information through telecommunications in vehicles. Among the applications of this technology are roadside-assistance services, navigation aides, vehicle tracking, stolen-vehicle recovery, and toll collection. Because these applications are growing rapidly, it has been suggested that this communications infrastructure might also be used to enhance the effectiveness of TOPIC devices by notifying authorities of alcohol violations in real-time.

The technical feasibility of sending data from these devices to authorities is not in doubt and the economic costs are modest. Monthly costs for a separate, dedicated communications link for a TOPIC device using the cell phone infrastructure would amount to about $10 to $12. The marginal cost of a link shared with roadside-assistance or vehicle-tracking services could be nil. The cost of adding a communications interface to a device could also be rather small in production quantities.

Our interviews with technology developers suggest that the major barrier to the use of telematics is the lack of interest and resources on the part of State and local governments in receiving and acting upon the information the interlocks might provide. They told us that relatively little use is being made of the information already being generated by interlocks. We believe that interlock programs are perceived as benefits for DUI offenders that allow them to regain driving privileges sooner than they would otherwise. The costs of the programs are borne entirely by the offenders in their monthly payments to the interlock service providers. Interlock vendors report that there are no appropriations for staff to examine data generated by interlocks and make appropriate responses. Hence, there is at present no demand for telematics for interlock programs as currently managed.

7.6 TARGET USER GROUPS

A 20-year period of development and refinement is typical of many technologies before they win market acceptance by a majority of potential buyers. Such an interval is suggested as a planning horizon for the technologies discussed in this paper. Automotive technologies, such as air conditioning, power steering, power braking, power windows, etc., were introduced in luxury cars shortly after World War II, but did not achieve 50 percent market penetration until the 1970s or 80s.

Since development efforts are likely to consume at least several years, there is no immediate need to identify prospective early users. However, there are several reasons to expect that fleets, especially government fleets, should be targeted for the early deployments of this technology.

The vehicular environment is characterized by shock, vibration, extremes of temperature, exposure to direct sunlight, etc. The destructive effects of these conditions usually delay wide-spread automotive application of a given technology by a decade or two beyond its stationary application, e.g., radios, air conditioning. The industry's well established approach to this problem is to introduce new technologies on a very limited scale, usually in the luxury market. Because the benefits of the technologies (e.g., all of the power-assist options) have been obvious, they could be marketed initially at high prices to high-end buyers. With small numbers of units sold at high prices, the industry can afford the high level of warranty claims and recalls associated with new technology. Early-adopting consumers tolerate its poor reliability because of the benefits it provides.

There is great doubt that ordinary buyers will opt for alcohol-impairment-prevention devices in the early years of their availability, because they are expected to be relatively expensive, relatively unreliable, and the crash-risk benefit depends mostly on other drivers, not one's own purchase decision. (Current BAIIDs are available for voluntary purchase at a cost of about $1,000, but sales are essentially nonexistent.) Yet without the prospect of substantial numbers of buyers, the industry may not invest in the technology development and refinement process.

To create a market for these devices, government intervention will almost surely be necessary. In its fleet-purchase decisions, the government can properly take into consideration the expected social benefits of a new technology. It can justify paying an initial premium for these devices and the operational costs associated with their malfunctions. Government employees using these vehicles will at least be paid for the time lost in dealing with the inevitable problems. The more problems that are found and fixed during the course of testing in government fleets, the smoother will be the introduction into private fleets and personal use.

Although devices engineered to function as primary interlocks could not be used unmodified as secondary interlocks, the same sensors could and should be tested in secondary applications. The great advantage of such testing is that secondary interlocks have data recorders and their users expect to visit service centers frequently for data downloading and verification that their equipment is in good working order. Any problems will be reported much faster and with much better documentation than would be the case for devices designed for non-offenders.

The next logical candidates for use of these devices are private fleet operators with particular safety concerns, such as common carriers of passengers, and hazmat carriers. Such firms have demonstrated interest in other impairment-prevention technologies related to fatigue and drugs, so it is likely they would be receptive to those related to alcohol.

7.7 COUNTERMEASURE ALTERNATIVES

Virtually all interlocks currently in use in North America do two things when a breath sample is taken that exceeds the set point:

- Prevent vehicle starting for some period of time; and
- Record the occurrences in memory, which is downloaded during periodic visits to service center. This data is then made available to authorities.

What a primary interlock should do is still under discussion. The full range of possibilities includes reporting violations to family, insurers, or law enforcement authorities, but these options are not being considered in the concept of operations at the sponsor's direction.

Among the possible responses to set-point exceedances are:
1. Immediate warnings to
    a. the driver,
    b. all vehicle occupants, and/or
    c. everyone in sight, by means of flashing hazard lights;
2. Preventing engine start for a period of time, which may
    a. be constant with each denied start, or
    b. allow a re-try a few minutes after the first failure, but require longer waits after subsequent failures;
3. Allow engine start, but lock transmission in "P" for some period of time as above;
4. Allow driving, but limit top speed to either
    a. a fixed limit for all roads, or
    b. a limit appropriate to the road being used, as determined by a GPS navigation system.

Warnings to inebriated drivers are regarded as ineffective by the interviewees who commented on the issue, and have been demonstrated as such in the research discussed in the following section. Warnings to other vehicle occupants could be helpful in some cases, but most of the time, either there are no other occupants or they are also drunk. Use of the hazard lights to warn everyone else and alert any police officers might be effective, but has not been tested.

An additional complication is that some jurisdictions require that an interlock have an emergency override switch so the vehicle can be used even if an exceedance has been sensed. The need for an override stems from the recognition that sensors can fail and that motorists occasionally drown or freeze to death in remote areas when their vehicles are disabled. Use of the override is reported to authorities and appropriate sanctions are applied, unless there is some good reason, such as a sensor malfunction or a life-or-death emergency situation. In a concept of operations without a data-reporting function, there is no obvious way to control the use of the override switch. If the switch is provided, a drunk driver could use it without penalty.

Given the constraints discussed in the preceding paragraphs, the viable countermeasure options are reduced to:
- Immobilizing the vehicle with an interlock so accurate and reliable that an override function is not required,
- Warning everyone in sight with the hazard flashers, and
- Restricting top speed.

These options could be used in combination and without an override. Circumvention of the hazard warning would be rather simple for most present-day vehicles with conventional wiring to the signal lights, but will likely become more difficult when these lights are bus-controlled.

7.7.1 Alcohol Warning (Qualitative or Binary)

No research appears to have been conducted on how drivers and passengers would respond to a display that indicates when alcohol has been detected in a vehicle. It is possible that an alcohol warning could bring family or other social pressure to bear upon the driver and reduce the likelihood of a reoccurrence. For this to occur, an alcohol warning should remain activated after alcohol is no longer detected.

7.7.2 BAC Warnings (Quantitative)

Although no research has been found on how alcohol-impaired drivers would respond to a vehicle BAC display, Nau, Van Houten, Rolider, and Jonah[90] obtained evidence suggesting that if a BAC level in excess of the per se limit were displayed, drivers would continue rather than stop. The Nau.[90] research team provided feedback on BrAC to patrons as they exited either of two Nova Scotia drinking establishments and whether the reading was over the per se limit of BAC = .08. Approximately half the drivers were impaired. They were then asked if they intended to drive home. If the reading exceeded the legal limit and the driver indicated an intention to drive, they were told not to drive. The research team then observed whether they drove. Nearly all (96 to 98%) of the drivers who had indicated an intention to drive did so.

Drivers who would receive vehicle feedback on BAC are similar to the drivers who received the feedback in the Nau [90] study except that they may already have begun to drive, and the in-vehicle BAC display could constantly warn the driver to stop. The authors offered a possible explanation for these findings "Feedback procedures may have been ineffective because the feedback was provided to drivers after they had become impaired" (p. 366). This would apply to in-vehicle warnings as well. Although the feedback was accompanied by increased enforcement, the latter may not have sufficiently increased the perceived likelihood of arrest for DUI, and as the authors noted, there were few alternatives to driving. In summary, this study suggests that in-vehicle BAC warnings would be ineffective, at least if implemented without complementary countermeasures such as visibly increased enforcement and acceptable alternatives to driving.

7.7.3 Impairment Warnings

Fairclough and van Winsum[91] studied how drivers respond to warnings derived from impaired behavior. Impairment was defined in terms of baseline performance collected during a 10-minute simulator familiarization drive. Normal driving extended to 40 percent above a person's baseline lane position variability and to 80 percent of the number of high-velocity steering corrections found in the person's baseline sample. The study imposed impairment by giving the drivers a monetary bonus in proportion to their times-on-task, up to 120 minutes. They lost the entire bonus if all four wheels departed the road. Off-road sound accompanied road departures. Subjects were instructed to drive until "they felt unable to continue at an acceptable level of performance" (p. 235) and were provided regular opportunities to stop along the course.

Impairment feedback consisted of three- and nine-level visual displays that were accompanied by three-level auditory displays. For example, the visual displays presented a green normal driving indication, a yellow warning if lane position variability rose 40 percent above the baseline values, and a red warning if it rose 70 percent above the baseline values. The three corresponding levels of auditory warning were no warning; "Warning. You are showing signs of impairment;" and "You are highly impaired. Take a break." Warnings were triggered upon either criterion exceeding its impairment criterion during each 30-second interval of the driving task.

Impairment feedback failed to influence drivers' decisions to stop or change speed. However, "it was apparent that participants choose to [quit] the journey principally on the basis of the [high velocity steering corrections]." In other words, the subjects were more sensitive to their own driving behavior than the criterion used to trigger the impairment warning. Apart from the requirement to keep all four wheels on the road in order to collect the time-on-task bonus, only a loss of the monetary incentive was provided to drivers who stopped. No incentives for speed control were provided in the study.

Impairment feedback influenced the specific driver behaviors corresponding to the measures that triggered the feedback: it reduced both lane position variability and high-velocity steering corrections. Feedback also succeeded in reducing related measures of lateral control including standard deviation of steering input and near-line crossings. Unlike near-line crossings, actual line crossings elicited feedback (off-road sound) in the control condition (the first 10 minutes of the driving test) as well as in the impairment condition, which may explain the lack of effect on this measure).

In summary, fatigued subjects modified their driving behavior in response to an in-vehicle warning system. Their behavior was sensitive to the events that triggered the feedback and to the incentive contingencies that were employed in the study. The results showed that when those contingencies (i.e., the monetary incentive to continue driving) conflicted with warning message content ("Take a break"), subjects responded in accordance with the incentive.

The study by Fairclough and van Winsum[91] was part of the research conducted by Project SAVE, discussed in Section 5.4.3. Any differences among the sources of impairment are regarded as relatively unimportant and fatigue is used only as a convenient source while drawing conclusions regarding impairment in general. However, sources of impairment have been associated with different effects. Even extremes of the same impairment dimension (i.e., temperature) have been shown to produce different effects on cognition.[92] To achieve confidence that these results generalize to alcohol-impaired drivers, one would need to replicate the study with alcohol-dosed subjects. The following section reviews evidence on whether alcohol impairment influences the effectiveness of feedback and incentives.

7.7.4 Performance and Incentive Feedback

The insurance industry reinforces safe driving through a system that assigns points for citations and crashes where the insured is at fault. Currently, some companies extend the principle of reinforcing safe driving by providing incentives for low-risk driving using periodic uploads of vehicle information such as speed to the company. It is conceivable that this principle could be further applied through in-vehicle displays associated with pay-as-you-go insurance. The question raised by this possibility is whether it would be an effective countermeasure against alcohol-impaired driving.

Some basic, laboratory research has been conducted on the effect of monetary incentives on mental processes involved in driving. "Automatic processes" are ones that occur with little or no effort, in contrast to "controlled processes," which require effortful thinking. In his review, Holloway ([93], pp. 42-43) distinguished the effects of alcohol on automatic and controlled mental processes that are required to drive: "Alcohol effects on automatic behaviors (e.g., turning) were seen only above 50 mg percent and in non-demanding situations, only at 70-80 mg percent, while clear effects could be seen at 30-40 mg percent in traffic situations requiring controlled processes (e.g., quickly-changing events) or having high social valence (e.g., heavy traffic, passengers, etc.)." Grattan-Miscio and Vogel-Sprott[94] investigated the influence of a small monetary incentive on alcohol-impaired performance of a laboratory task designed to separate automatic from controlled processes. Word stems presented in green were to be completed with familiar words that had just been shown, permitting automatic processes to facilitate controlled processes. For example, TO--- might be shown after the subjects saw the word TOPIC in a preceding list. Word stems presented in red were to be completed with words that had not been shown requiring controlled processes to counter the automatic process that would complete a word stem incorrectly with the familiar word. The alcohol dose used produced BAC = .07. Compared to placebo, alcohol was shown to substantially reduce the influence of controlled, but not automatic processing. A small monetary incentive eliminated this reduction in controlled processing. This research suggests that insurance discounts or other incentives (and possibly disincentives) triggered by the detection of an alcohol signature might be an effective countermeasure against alcohol impairment of controlled processes involved in driving. .

A second study examined the influence of a small monetary incentive on how alcohol (BAC = .07 to .08) affects the capacity and accessibility of information in working memory (WM).[44] The experimental paradigm showed a set of letters to remember and then probed with a letter that was either in the memory set or not. This procedure has been shown to cause the subject to mentally "scan" the memory set and thus to test the use of working memory. A small incentive was found to "counteract the slowing effect of alcohol on the rate of scanning and RT [response time] when WM ... was taxed to capacity ... [and] to restore these aspects of WM to the drug-free level shown under a placebo...." (p. 194). The incentive was implemented in the procedure by telling the subject whether or not their performance was better than their pretreatment baseline performance following each block of test trials.

The amount of the incentive was quite low in the preceding studies (a maximum of $2 in the latter and comparable rewards in the former) hence its effects can be taken as support for an effect of performance feedback on alcohol-impaired task performance. While these findings do not apply directly to the driving task as a whole, they do suggest that individuals under the influence of alcohol near a per se BAC limit of .08 would respond to performance feedback and/or minor incentives. This in turn suggests that the effect of performance and incentive feedback that was found for improving fatigue-impaired driving[91] would generalize to alcohol impairment.

7.7.5 Crash Avoidance Warnings

The preceding research indicated that alcohol-impaired performance was improved by incentive feedback that *followed* task performance. However, unlike the impairment warnings that Fairclough and van Winsum[91] found to be effective against fatigue, crash avoidance warnings, for example, headway and lane departure warnings, are *predictive* cues to impending hazards. For this reason it is important to determine whether alcohol impairment would affect a driver's sensitivity to cues that predict events that are about to occur. Fillmore[95] demonstrated that the sensitivity of performance to predictive information increases with alcohol dose (placebo, BACs = .045 and .065). He demonstrated this with a "cued go/no-go task" in which the orientation of a rectangular frame predicted with 80 percent validity whether the frame would be filled with a green or blue target. Subjects were to "go" (i.e., respond to the green target by pressing a key as quickly as possible) and to the blue target by a "no-go" response (i.e., by inhibiting the key press). Alcohol increased the effectiveness of the predictive frame.

If they generalize, these laboratory studies would mean that alcohol-impaired drivers may be more sensitive to incentives and predictive information than non-impaired drivers. They suggest that alcohol impairment may not decrease the effectiveness of in-vehicle warnings; it could even increase their effectiveness. However, in-vehicle warnings may do more than present incentives and driving performance feedback. While they increase motivation or awareness of lane position, for example, they may also distract drivers from primary aspects of the driving task.

7.7.6 Crash-Avoidance Technologies

7.7.6.1 Description

Crash-avoidance technologies have been the subject of major development programs on the part of the world's major automobile manufacturers and tier-one suppliers in recent years. Three classes of crash-avoidance technology have begun to appear in some vehicles. These technologies are designed to reduce crash risk for all drivers, but they may be especially beneficial to impaired drivers, simply because the latter make more errors. However, there is at present no experimental data to show how alcohol-impaired drivers will respond to warnings and/or autonomous control actions from these new systems. The functions of these systems are summarized as follows:

Electronic Stability Control Systems

ESCs are comprised of a steering-angle sensor, yaw-rate and lateral-acceleration sensors, wheel-speed sensors, brake-pressure sensors, a hydraulic modulator, and a microprocessor-based controller. These components sense the onset of conditions leading to skidding and rollovers, and automatically apply braking forces to the appropriate wheels as necessary. Every major manufacturer is now offering these systems on selected models and market penetration is projected to reach nearly 25 percent in the 2007 model year. These systems are typically priced at $300 to $2,000 as options, but are standard on some luxury cars.

Adaptive Cruise Control/ Forward Collision Warning Systems

FCW systems employ either microwave radar or laser infrared sensors (LIDAR) to detect and measure closing rate to obstacles (mainly other vehicles) in the path of a vehicle. Some use GPS to refine control algorithms. Whenever an excessive closing rate is detected the controller generates a warning signal, reduces speed, and may release the throttle completely. Gentle braking may be applied automatically, but hard braking is left to the driver. These systems are available on several luxury models now either as standard equipment or in an option package priced at $2,000 to $3,000.

Lane Departure Warning Systems

LDWs use an array of video cameras and sophisticated image-processing software to recognize lane markings and road edges and to measure the distance and closing rate between a vehicle and the limits of its lane. Additional cameras sense vehicles in adjacent lanes, and some systems incorporate GPS-assisted curve warnings. Warnings are provided when a driver appears to be approaching a boundary unless the driver has indicated a deliberate intent to cross the boundary through use of the turn signal. Warnings of vehicles in blind spots may also be given. These systems are currently available in a very small number of models, typically as a $4,000+ option.

7.7.6.2 Performance

Performance standards for these systems have not yet been developed. The complexity of the measurements they perform and the diversity of system responses makes this task inherently much more complex than for other automotive systems. The best indication that that they work well enough to justify their cost is simply that a substantial and growing proportion of new-vehicle buyers are willing to pay for them.

A major issue in performance of crash avoidance systems is that of false alarms. Unless the false-alarm rate is kept very low, consumers tend to ignore warnings, disable the devices, and avoid purchasing them in the first place. For automotive warning systems, this issue is being addressed through the development of sensors to determine whether the driver is looking at the road,[15] and to suppress warnings whenever the driver's attention is focused there.

7.7.7 Potential for Distraction from In-Vehicle Warnings

Iudice et al.[96] examined the effect of distraction on alcohol-impaired drivers (BAC = .05) using a simulation that required subjects to respond to events including traffic lights, crossing roads, pedestrian crossings, and oncoming and preceding traffic. Subjects were instructed to drive as they would normally. Concurrent and continuous divided attention tasks were presented over a hands-free cell phone including backwards spelling and counting, and arithmetic. Compared to a sober, distracted baseline condition, alcohol-impaired subjects completed the 15-km course in significantly less time (although with no speed exceedances). Time-to-collision with preceding traffic showed a sizable although non-significant reduction (2.4 seconds versus 4.7 seconds in the baseline condition). These results provide evidence of higher speeds and a trend toward more aggressive driving when the drivers were distracted and alcohol-impaired than when they were only distracted.

The alcohol level that Iudice et al.[96] employed is the per se limit in many countries, but less than the limit required in the United States and much less than the level found in drivers who were involved in fatal crashes.[33] In addition, they did not provide a placebo and undistracted-driver baseline for the comparison of alcohol and distraction so it was not possible to determine whether the alcohol-impaired, distracted drivers would have shown a significant effect on time-to-collision. Rakauskas and Ward[97] used BAC = .08 for their alcohol conditions as well as an

---

[15] The SAVE-IT project, sponsored by NHTSA, is developing a way to monitor drivers' attentiveness to the road and suppress warnings if attentiveness is detected.

undistracted, unimpaired baseline. The simulation task was to maintain a safe headway while the lead vehicle accelerated and decelerated. In-vehicle tasks provided auditory instructions and required visual attention to manipulate the controls of a Compaq iPAQ computer located on the dashboard. Hands-free cell phone tasks distracted the driver in other conditions with auditory requests to orally repeat a sentence, solve a verbal puzzle or discuss a specified topic. Although several headway indices were obtained, none of the results supported the hypothesis that the subjects would become more sensitive to distraction. However, further data analysis revealed an increase in lane position variability and the application of more steering power when alcohol-impaired subjects were distracted than when unimpaired drivers were distracted.[98] The studies by Iudice et al.[96] and Rakauskas and Ward[97, 98] indicate that alcohol-impaired drivers are more subject to distraction by tasks extraneous to driving than unimpaired drivers. More specific studies are needed to determine whether similar results would be found with impairment warnings or crash avoidance warnings.

Victor et al.[67] found that an in-vehicle task requiring vision caused drivers to look less at the road ahead and more at the visual display. When a driver's gaze returned to the road from the visual display, it returned to the center of the road. An auditory task also concentrated dwell on the road center and away from the road periphery, an effect that Recarte and Nunes[99, 100] found to result in less awareness of events occurring at roadside. These studies may have design implications for the presentation of warnings and performance feedback to alcohol-impaired drivers. There maybe design implications if, for example, performance feedback shows a reduction in gazes to the side of the road similar to in-vehicle tasks and if this effect is additive with the "tunnel vision" effect of alcohol (Bel, 1969, cited in [101]).

In summary, these studies show a potential cause for concern regarding the presentation of information such as impairment, crash avoidance warnings, and incentives to drivers. The possibility exists of a tradeoff between the benefits of the information with an increase in distraction that may increase the crash risk of impaired drivers.

7.7.8 Acceptability of Incentives and Driving Performance Feedback

Researchers have explored the acceptability of using in-vehicle feedback on driving performance. Roetting, Huang, McDevitt, and Melton[102] conducted focus groups with truck drivers and other experts from the trucking industry followed by a survey completed by a sample of long-haul truck drivers.[103] Results from the focus groups indicated that the participants desired "specific, constructive, respectful and individualized" feedback, particularly meaningful positive feedback "accompanied by signs of recognition, like a bonus or reward" ([102], p. 282). However negative feedback, such as collision warnings, was recognized as sometimes necessary. While feedback from a human was preferred, well designed feedback from technology was considered acceptable, particularly when accompanied by human feedback. Perceived benefits included lower operating costs due to fewer crashes and lower insurance rates.

Focus groups were "consistently" concerned about privacy issues and "would not feel comfortable with being watched by technology" ([102], p. 283). In the later survey, 65 percent of the long-haul truck driver sample indicated a concern about the potential misuse of data collected by in-vehicle technology. On the other hand, "the greatest perceived benefit of technology was the use of the recorded data in defending the driver if he or she would be involved in an incident" ([103], p. 289). Another concern found in the focus groups was with over-reliance on technology to tell them when they are not driving safely. This finding was confirmed by 52 percent of the survey respondents.

The survey also examined the acceptability of various forms of feedback. A majority (56%) preferred to receive feedback immediately after an event, and half preferred feedback upon request rather than presented automatically. A majority (57%) also said they would not want to receive feedback from a computerized voice, whereas 20 percent said that they would want to receive feedback in this way. Feedback from a computerized voice was clearly preferred to a visual display: 47 percent versus 20 percent.

8. A CONCEPT OF OPERATIONS FOR A TECHNOLOGY UNDER DEVELOPMENT: PRIMARY INTERLOCK USING TISSUE SPECTROSCOPY

8.1 OBJECTIVES AND TARGET USERS

The purpose of this Concept of Operations is to explain how BAC detectors based on NIR tissue spectroscopy might be applied as primary ignition interlocks to new vehicles in the North American market. A critical assumption in this concept is that such use will not be mandated by either State or national governments – rather that these interlocks must have attributes that will attract substantial numbers of new-vehicle buyers to order them voluntarily. It is possible that these devices will be ordered by so many buyers that they will become standard equipment, but whether and when that happens will be left to the vehicle manufacturers.

A second critical assumption is that the proposed concept of operations entails no data recording or reporting. So doing would facilitate the offering of risk-based insurance discounts and provide useful information to those managing the treatment of problem drinkers, but the privacy issues and marketing negatives may to be so formidable as to exclude the concept.

The ultimate target users are the entire population of motor-vehicle operators, but the process of reaching them is likely to take decades. Like most other new vehicular technologies, these interlocks are likely to be sold first in more expensive vehicles. Such buyers can more easily afford the initial costs and perceive greater savings from accidents avoided than low-end customers, but are more demanding as to ease of use, reliability, and maintenance burden.

8.2 PHYSICAL CHARACTERISTICS AND PERFORMANCE SPECIFICATIONS

Designing and producing an NIR alcohol interlock that can appeal to voluntary buyers poses enormous engineering and economic challenges. Compared with existing prototypes, the device must be reduced nearly three orders of magnitude in volume and at least one in cost while maintaining accuracy, reducing measurement time, and becoming maintenance-free. These reductions are similar in scale to those of cell phones from their introduction in the early 1980s to the present. However, cell phones sold in substantial numbers even when they weighed three pounds and cost a thousand dollars. This sales revenue provided funds for investment in subsequent generations of smaller, cheaper phones appealing to ever-larger groups of buyers. The difficulty for the NIR alcohol interlock is that there may not be any voluntary buyers until the device can be made much smaller and cheaper than it is now.

Low-volume electronic products, such as current alcohol interlocks, are produced using many off-the-shelf parts and a largely manual assembly process. Achieving very small physical size and low cost requires a large investment in both product and process technology. The source of funds for such investment is unclear, but must be recognized as a major barrier to the implementation of this concept of operations. Fortunately, most of the needed technologies for sensors, processors, and software are being developed for other applications, especially medical instrumentation.

The design of these devices must address the issues discussed in the following subsections.

8.2.1 Accuracy

According to Ridder et al.[52, 53] the demonstrated accuracy of the TruTouch prototype is already more than adequate, at least under the conditions tested thus far. Its RMS prediction error is equivalent to about .005 of actual BAC, which is about one-third of the error for evidential breath analyzers. However, these statements are apparently based on samples taken mostly from subjects with stable or declining BACs.

Currently unknown is the extent to which chemicals other than ethanol may be present in some drivers' bodies and which modify the observed spectra in the 4100-to-4300 wave-number range. The prevalence of such confounds is only a matter of speculation at present. A period of some years of testing will be needed to clarify these questions. Much of this testing is likely to occur in clinical trials where NIR spectroscopy will be used to study a variety of blood analyses.

Aside from TruTouch, none of the other proposed ideas for NIR-spectroscopic detectors have been developed to the point that they have been tested for accuracy in human subjects.

8.2.2 Set Point

The trigger level or set point for future primary interlocks is open to discussion. Current interlocks for offenders are set to trigger at BAC values ranging from .02 to .04 in the United States. Many States specify .025 (the value NHTSA requires for calibration tests). Elsewhere in the world, set points range from 0 to .05.

The rationale for these low set-point values is that offenders should not drink at all when they anticipate driving. (They are non-zero to eliminate false positives caused by detector noise and error.) This concept makes sense to the law-enforcement officials and offender counselors we interviewed, because most current interlock users have serious alcohol problems with little or no ability to limit their drinking once they begin.

However, the argument for a low set point does not apply to vast majority of drivers. Furthermore, the suggestion of effectively eliminating moderate social drinking from most of the occasions in which it currently occurs is certain to engender strong opposition. While interlocks with low set points could be sold to teetotalers and to commercial fleets, marketing them to the large proportion of new vehicle buyers who drink moderately would be extremely difficult. The expected insurance discount (the principal inducement for buying an interlock and amounting to no more than $200 per year) would not be nearly large enough to offset the cost and inconvenience of taxis or other alternate transportation for all of the occasions on which social drinking normally occurs over the course of a year.

Given that statutory limits for legal driving are now set at .08 throughout the United States, it is difficult to make the case for any other value as the set point for primary interlocks. That value is high enough to allow moderate social drinking, but low enough to block nearly all of the driving associated with fatal accidents, and by definition protects the driver from committing a DUI offence.

8.2.3 Latency

Alcohol concentration in tissue lags behind that of blood during the uptake phase by about 30 minutes, but tracks very closely during excretion. Sparse available data suggest that true BAC may be underestimated by about 40 percent during uptake, but there is virtually no latency problem once BAC has peaked. Conversely, in BrAC measurements, the larger errors occur while true BAC is declining. Presumably, impaired driving occurs during the latter period as well.

The spectroscopic measurement itself is nearly instantaneous and introduces no additional latency. There is no requirement for a specification, merely recognition of the inherent latency in tissue uptake.

8.2.4 Form Factor

Ease-of-use considerations imply that the detector must be embedded in something that the user must be touching in the normal course of driving or preparing to drive, so that the test causes no delay and is completely passive. Two obvious candidates are the steering wheel and the key fob. Each has its advantages:

- **Steering Wheel.** Embedding an NIR alcohol sensor in the steering wheel has several obvious advantages. At least one of the driver's hands is almost always in contact with it, which simplifies rolling retests. Only the sensor itself needs to be really small; the rest of the electronics can be located wherever convenient. Power and CANbus connections are already present in the steering wheel hub of many cars. Power consumption and battery life would not be matters of concern. To avoid compromising the structural integrity of the steering wheel, the sensor volume probably needs to be reduced to something on the order of two cubic centimeters or less. The rest of the circuit functions would most likely be carried out in one of the microprocessors on the vehicle's LAN.

- **Key Fob**. If the device can be made small enough and if sufficient battery life can be achieved, a key fob version becomes a viable concept. The principal advantage of this packaging is that it allows a driver who has been drinking to determine whether the driver is under the limit without actually getting into the vehicle. This offers obvious advantages in terms on convenience and avoidance of embarrassment. The sensor, processor, power source, and wireless interface to the vehicle would be constrained to a maximum volume of a few cubic centimeters – comparable to what has already been achieved with solid-state BrAC analyzers embedded in cell phones.

8.2.5 User Friendliness

There is complete agreement that to win consumer acceptance, primary interlocks must epitomize user friendliness -- to wit: no time, no effort, and no thought. While no technology can achieve these ideals, tissue spectroscopy appears to hold the best prospect for approaching them. Current prototypes of evidential testers require measurement times of almost 30 seconds and placement of the subject's entire forearm in the test fixture. There is a possibility that the sensor can be reduced to a size that can be embedded in the steering wheel, probably at the 10-and 2-o'clock positions, and that reading time can be cut to a few seconds. Some type of indicator must be provided to prompt drivers to place their hands on the wheel (or other sensing surface) each time a reading is required.

The interlock must recognize all possible operator errors, such as, wearing gloves, wrong placement of hands, skin covered with a material that interferes with spectroscopy, etc, and provide appropriate instructions to the driver on a dashboard display.

8.2.6 Feedback

All current interlocks include a digital display that shows a BAC reading for each breath sample taken. This feature provides guidance to drivers about how close they are to the set point and how rapidly they metabolize alcohol. It should help to avoid the embarrassment of exceeding the set point and clearly contributes to user friendliness. Should the sensor start to go out of calibration, abnormal readings can alert the driver that service is needed before the problem causes denied starts. The display should be retained in all primary interlock designs.

8.2.7 Retesting Intervals

Specifications for two different types of retest intervals must be determined. The first is the interval after a failed pre-start test and before another test is permitted. In current interlocks for offenders, these intervals are set at the discretion of the manufacturer. Different choices lead to different rates of denied starts. One approach that makes sense for BrAC interlocks is to allow the first retest after a short period of five minutes, which is often sufficient to allow the effects of mouth alcohol to dissipate. After a second failure the retest interval increases to 30 minutes to prevent a drunk from recording a large number of denied starts. (Some States impose sanctions on drivers who exceed a specified number of denied starts per month.)

Because the NIR interlock is not affected by mouth alcohol, the experience with BrAC interlocks does not provide much guidance on this point. Since there is no reporting in the contemplated implementation of primary interlocks, there appear to be no disadvantages to allowing frequent retests – say at five-minute intervals. Research is needed for further guidance on this issue.

The other aspect of retest timing is the interval between samples once driving has begun. The sampling procedure for breath devices is so distracting that it the recommendation is to perform it with the vehicle stopped. Hence, the average of the random sampling interval is set to around 30 minutes. However, with tissue spectroscopy, the sampling procedure should make no demands on the driver other than that the driver place both hands on the wheel (or one finger on a sensor in some other location, such as the key fob). A short retest interval is thus feasible. Furthermore, a shorter retest interval makes circumvention by having someone other than the driver touch the sensor(s) more difficult. Hence, a retest interval in the 10- to 15-minute range is recommended.

8.2.8 Environment

Operation in a motor vehicle exposes any device to a harsh environment including vibration, temperature extremes, sunlight, numerous sources of electromagnetic interference, chemical vapors, dirt, and dust. While there is no reason to believe that any of these conditions present insurmountable challenges, it is instructive to note that the automotive versions of nearly all electronic products have not appeared on the market until a few years after the stationary and personal portable versions of the same products. The delay results from both the time required to solve the environmental problems and the time required to integrate the new product into the interior design of vehicles.

Most of the environmental specifications for an NIR interlock would be similar or identical to those of current BrAC interlocks[30], which is currently being revised and updated. Because of its different operating principle, an NIR interlock will require additional specifications related to its ability to withstand and function properly in the presence of the following environmental challenges:

- Strong direct sunlight (100,000 lux), which could interfere with measurements or even damage a detector if not properly designed;
- Presence of various common types of dirt, oil, cosmetics, etc., on the hands of drivers; and
- Wear and abrasion of the lens covering the detector.

Appropriate performance tests with respect to these and other as-yet-unknown environmental issues remain to be developed and validated.

8.2.9 Reliability and Maintenance Burden

There is consensus among all stakeholders that a primary interlock must be highly reliable and impose minimal maintenance burdens on users. Essentially, it must be as reliable as other items of automotive electronics and should be designed and manufactured in a process that conforms to ISO/TS 16949. Scheduled maintenance should occur no more frequently than for other vehicular systems – typically at 30,000-mile intervals.

8.2.10 Verification That Tissue Sample Tested Belongs to Driver

Current BrAC interlocks are somewhat vulnerable to circumvention by means of another person providing the breath samples. (Beside being illegal, this method is often difficult in practice because it takes practice to blow an acceptable sample, and the other occupants are often absent, or also drunk.) A user-friendly NIR interlock might be easier to circumvent, unless it explicitly includes some means to ensure that the person who is sitting in the driver's seat is the person who is being tested for BAC.

The choice of means is left to vendors, but may involve: (1) placement of two sensors at locations that would be difficult for anyone other than the driver to reach simultaneously, (2) frequent rolling retests, or (3) confirmation that the person being tested is among the small group of authorized users by matching tissue-scan images. (The latter method is also part of the theft-prevention aspect expected to be marketed with the interlock.)

It is possible that unscrupulous individuals or firms may produce and sell materials that mimic the spectroscopic characteristics of human tissue with no ethanol in the 4100-4300 wave number range. Hence, it may be necessary to scan and analyze a wider portion of the spectrum so as to include additional tests to verify that the sample being scanned is in fact live human tissue.

8.2.11 Anti-Circumvention Features

Given that this concept of operations entails no data reporting system nor regular visits to an interlock service center, circumvention may be far more difficult to prevent than is the case for current offender interlocks. If the interlock were implemented by means of a relay in the ignition circuit (like current devices) circumvention would require only a jumper wire across the relay, the installation of which is well within the abilities of a large portion of drivers. The presence of such a jumper would be evident to an inspector, but at present there is no legal requirement for such inspection, nor sanction for having the jumper except for participants in offender programs. Circumvention would be still more difficult to detect if a driver unplugged the relay and replaced it with a jumper plug, but reinstalled the relay during his annual inspections.

Integration of the interlock with the vehicle's computer system is strongly recommended because it eliminates the easy methods of circumvention available to any "shade tree" mechanic. However, a black market could develop for sophisticated circumvention devices, such as those that exist for cable descramblers or pirated software. How to design a system to minimize such vulnerabilities is far beyond the scope of this report, but the importance of the issue must be recognized. It is likely that some means of checking the interlock system for hacked software or hardware should be incorporated in annual inspections.

8.2.12 Vehicle Interface

For reasons of both economy and anti-circumvention, integration of the NIR interlock into the vehicle's computer system is recommended over an implementation as a separate device with hardwired connections to the vehicle's ignition and signal-light circuits. The extent to which the computationally intensive task of running the partial-least-squares model can be handled by the vehicle's computer is unknown at present. Existing dashboard displays and annunciators should be used to prompt drivers to place their hands on sensors and show BAC levels.

8.2.13 Economic Constraints

One important way of deciding what concepts of operation are viable is a consideration of the economics of drunk driving. The *Impaired Driving in the United States* estimates prepared by PIRE [104] for the year 2000 show that the total social costs of alcohol-related crashes in that year were $114.3 billion, of which $51.1 billion were monetary, while the remainder were for pain, suffering, loss of companionship, etc. Given a total vehicle population of about 221 million in 2000, this works out to $517 per vehicle (total cost) or $231 (monetary cost) per vehicle per year. These numbers might be considered upper bounds for the justifiable annual costs in 2000 of an alcohol-crash-prevention concept that is universally applied and 100-percent effective.

To adjust these values to the present, they should be increased by at least the change in the GDP-deflator (+14 percent) and the change in real GDP per capita (+8 percent). These two factors (1.14 * 1.08 = 1.23) raise the estimates for 2006 to $654 (total) and $284 (monetary) per vehicle. These numbers are conservative, because some crash-related costs (medical treatment, vehicle repairs, etc.) are rising faster than overall prices.

The average monetary costs, $284 per vehicle, are a reasonable upper limit for how an insurance company might value the addition of a 100-percent-effective crash-prevention concept to its customer's vehicle, assuming that the risks are uniformly distributed across all customers. The actual discount offered to the customer would likely be smaller because no system is 100-percent effective and some alcohol-related crashes would have occurred even if the driver had been sober. For purposes of illustration we, optimistically assume the savings and discount might be as large as $200 per year for the average vehicle.

New-car buyers who do not plan to drive drunk and who base their decisions solely on economic considerations should be willing to pay for the technology an amount up to the discounted present value of the stream of insurance-premium savings. Over a range of discount rates from 6 percent to 10 percent, the value of a $200-per-year insurance savings ranges from roughly $1,700 to $,1400, assuming a 12-year vehicle life. To the extent that the purchase of an interlock entails any inconvenience to drivers, their valuation of it will, of course, be lowered.

Such sums are more than sufficient to cover the initial costs of a number of technologies. The installed cost of present-generation BrAC interlocks is about $1,000, for example. However, routine maintenance costs might wipe out the economic advantage to buyers. Current offender interlocks require so much service (data collection and reporting, recalibration, etc.) that their annual cost exceeds $900.

For concepts to be applied universally and voluntarily, the following constraints apply:
1. The initial incremental cost of a crash-prevention concept to a vehicle buyer should not exceed $1,500 (2006 dollars).

2. Maintenance costs must be very low – as low as car radios, for example.
3. Achieving this level of reliability and durability is likely to require a period of some years between the time the technology is ready for some markets (e.g. clinical monitors) and the time it is ready for the automotive mass market.
4. Insurance companies will require some assurance that the device is installed and working properly to prevent drunk driving in order to offer and maintain the discount. Legal procedures must be devised to provide certification that the device was installed when the vehicle was sold and verification of its continuing functionality at annual inspections.
5. The above implies that a viable technology must include effective measures to prevent circumvention by any means that would not be recognized during annual inspections as well as a fast, low-cost method to test its functionality and accuracy at such inspections.

There is a near consensus among stakeholders that insurance discounts are a critical element in the primary interlock concept. The notable exception to the consensus was expressed by the insurance-industry representatives. Though not denying the possibility of substantial discounts someday, the insurers raise the following objections:

- Discounts should be based on actual reductions in claim losses – not projections from simulator studies or small field tests. The change in overall dollar claims may be substantially different from the change in fatalities or serious accidents. This aspect has usually been neglected in previous studies. The experience to date with electronic stability control systems bears out this point. Overall claim savings are much smaller than the percentage change in fatal accidents.
- Unlike other crash-reduction technologies, the risks from impaired drivers are NOT uniformly distributed. Customers who never drive after heavy drinking will not experience much reduction in crash risk, and insurers realize they have nothing to gain from inducing abstainers to install interlocks.
- Early decisions about offering discounts will be made by companies individually and will depend on such considerations as:
    o Whether a particular firm wants to try to select low-risk clients in the first place. (Some do; others make money selling to high-risk drivers.)
    o The claims-loss experience in Europe with primary interlocks, which is as yet undocumented.
    o The claims-loss experience in the North America with interlocks for offenders.
    o The results of field operational tests of primary interlocks.

Early discounts may be modest and limited. Only after some period of favorable claims experience are the discounts likely to approximate the actual savings brought about by the presence of primary interlocks.

8.2.14 Technology Bundling

An effective approach to marketing an unfamiliar option to new vehicle buyers is to bundle it with a package of options that have already won consumer acceptance. This tactic is recommended for primary interlocks.

NIR spectroscopy is said to be capable of identifying structural features in tissue that are at least as good as fingerprints. Thus it can determine whether a particular driver belongs to the set of drivers who have been previously authorized to use a given vehicle. Hence, bundling the interlock with theft-prevention options is obvious.

Experience with ABS suggests that the response of drunk drivers differs from sober drivers. [105] This may also hold for various crash-avoidance technologies, although no research has yet been conduct to provide guidance. In the event that a need to modify the response of crash-avoidance systems is demonstrated for drunk drivers, the bundling of NIR monitors with them is indicated.

9. CONCLUSIONS

9.1 TECHNOLOGIES IN USE: SECONDARY INTERLOCKS

Secondary interlocks have demonstrated effectiveness in reducing DUI-recidivism, but fewer than 8 percent of offenders are actually using them. The rather limited data available suggests that the crash rate of interlock users is similar to that of the non-offender population. Many issues contribute to their low rate of use ranging from insufficient judicial awareness of their potential to concerns about cost.

Alcohol-impaired drivers with a BAC ≥ 0.08 face about a 1 percent risk of a DUI arrest. Repeat offenders constitute about one third of the 1,014,000 arrests for DUIs nationally, and they tend to have more driving violations. The self-reported recall data on the incidence of driving under the influence suggests that drivers gamble on the low levels of enforcement. More aggressive enforcement using more police could increase the risk of arrest.

It is estimated that widespread installation of secondary interlocks, BAIIDs, in 100 percent of the vehicles driven by first, as well as repeat, DUI offenders, in commercial vehicles, as well as in all vehicles driven by drivers under age 21, would decrease drunk driving crash deaths 25 to 40 percent and would have prevented an estimated 3,000 to 5,000 of the 12,677 fatalities in 2004.

9.2 TECHNOLOGIES UNDER DEVELOPMENT: PRIMARY INTERLOCKS

There is no technology ready for near-term use as a primary interlock. Tissue spectroscopy has the most promising characteristics, but must be reduced by three orders of magnitude in size, reduced one order of magnitude in cost, and re-designed to work on palms and fingers.

All technologies are highly vulnerable to circumvention unless an infrastructure is established that permits devices to report circumvention to authorities, who can and will impose appropriate sanctions. Alternatively, an interlock might be developed that is inherently invulnerable to circumvention through:

- Secure integration with the vehicle's engine control computer,
- Inclusion of additional test features to verify that the sample being tested is the driver's skin, and
- Capability to perform accurately and reliably throughout the life of the vehicle.

If potential new technologies are found that are suitable and unobtrusive enough to be fabricated as primary interlocks, it is estimated that universal adoption of primary interlock devices would eliminate 30 percent of the traffic fatalities, or 12,677 of the 42,636 fatalities incurred in 2004.

9.3 TECHNOLOGIES UNDER DEVELOPMENT: VEHICLE-BASED IMPAIRMENT MONITORS

Behavioral indicators offer the potential to establish personal and/or general baselines for defining impairment that could form the basis for control functions to reduce crash risk beyond what is possible to achieve with primary interlocks.

General baselines are more challenging to achieve, but, in theory, they offer the potential for more reduction of crash risk because, like police officers on the road, they would use an absolute definition of impairment, whereas, personal baselines use a definition that is relative to whatever level of skill a driver normally exhibits. Behavioral impairment monitoring is not a substitute for interlocks that prevent impaired driving before it starts, and interlocks are not a substitute for systems based on behavioral indicators which could respond to impaired driving at BACs less than the per se limit in combination with fatigue or other sources of impairment.

It may be difficult to untangle alcohol impairment from drug use and fatigue using behavioral indicators, and we have found no results that address this possibility. Although this was an objective of Project SAVE, results bearing on this issue have not been disclosed. However, it should be noted that it is inherently more difficult to distinguish the source than to identify impairment. Not only must the detector identify impairment, but it must also identity its source. To accomplish this, the detector must be capable of discriminating among all sources of interest. Increasing the categories of impairment that must be separated makes it harder to get a "hit." Not only might the identification of an alcohol-specific "signature" be difficult to achieve, its use would reduce impairment-related crashes less than systems that detect driving that has been degraded by any source of impairment. However, the development of countermeasures specific to alcohol (e.g., a BAC alert, notification of law enforcement, or CWS adjustments for specifically alcohol-impaired drivers) may justify what currently appears to be a difficult program of research leading toward the identification of an alcohol-specific behavioral signature.

10. RECOMMENDATIONS FOR TECHNOLOGIES IN USE

The following recommendations relate to the use of BAIIDs as secondary interlocks. Some of these recommendations also apply to primary interlocks. These recommendations have been suggested by one or more of the numerous subject-matter experts interviewed, as well as from insights gained from the research literature.

Increase the number of offenders using BAIIDs to reduce DUI recidivism.

- Consider the advantages of an automatic, administrative process that requires interlocks for DUI offenders.
- Identify effective programs for first offenders and explore why some programs for first offenders fail to reduce recidivism significantly.
- Consider the application of interlocks to first offenders and monitor the experience in New Mexico and Washington for high-BAC first offenders.
- Consider allowing repeat DUI offenders who install BAIIDs to reinstate their driver licenses early after serving their suspensions.
- Consider encouraging more repeat offenders to install BAIID interlocks in order to regain valid driving privileges
- Provide accessible information for enforcement authorities such as judges about BAIIDs and how they work.
- Recommend that the duration of BAIID use depend on the offender's performance rather than a specific date for its removal. Analyze efficacy of State policies on use of BAIIDs, i.e., some States are starting to make the duration of use contingent on driving without any denied starts.
- Review State laws to identify conflicting laws regarding interlocks which make judges reluctant to impose them.
- Support strong sanctions or disincentives against circumventing the BAIID by driving a vehicle that is not equipped with an interlock.

Make more use of BAIID data

- Require BAIID providers to send information to the State and standardize procedures for formatting and transmitting data.
- Recommend that States track the data returned to them by the BAIID makers and provide templates for tracking procedures.
- Require that BAIID usage information to be sent to treatment programs.
- Analyze the variability in rates of denied starts across jurisdictions to improve selection criteria for inclusion in interlock programs.
- Use interlock data to control the duration of interlock use and/or to provide feedback for counseling programs.

- Consider expanding access to BAIID data such as alerting others, e.g., employers, family, when a breath test was failed.
- Investigate where the data on failed starts goes. Does it go to the courts? Treatment programs?
- Develop common standards for the data expected from expanded interlock use.

Ease BAIID and IID installation issues

- Provide NHTSA standards for the OB-II interface for future products such as new BAIID models. This would help manufacturers decide what to implement as the industry is moving to wireless data collection and remote calibration to cope with a larger market for interlocks.
- Consider whether BAIIDs might be implemented using a Bluetooth link through the "Accessory" node of the vehicle's local area network.

Improve BAIID level of service

- Consider whether interlock distributors need a certified level of competence.
- Ensure that the certification process proceeds quickly and that the States do not have to repeat new interlock conformance to NHTSA standards tests, which delays the introduction of new products.

Lessen the economic burden of mandating BAIIDs

- Recommend State purchase of BAIIDs as a way to increase their use. BAIIDs could revert to the State for reuse upon completion of the sentence.
- Consider requiring the offender to pay up front for a BAIID and be reimbursed upon completion of the program.
- Conduct a field test to demonstrate that BAIIDs reduce crashes significantly. Insured-loss claim data should be included in the test and a commercial fleet could be used. Because it has an administrative program and a single service vendor, the province of Alberta is a good place to conduct research to understand impaired driving and the effects of BAIIDs while minimizing exogenous factors.
- Monitor the results of the campaign in Sweden to get insurers to offer discounts on interlock-equipped cars.

Enhance the usability of BAIIDs and IIDs

- Eliminate distraction problems with BAIIDs. Many units beep loudly after one minute, causing many drivers to perform retests while in motion.

Address the remaining gaps in BAIID evaluation research

- Compare exposure-weighted crash rates for interlock users with corresponding rates for a matched group of nonusers. This is the ideal research but no such study has taken place because of the methodological difficulties, privacy issues, and expense.

- Carry out longitudinal studies of interlock users. This would be informative but only a few studies have been performed, and to our knowledge, none include data from the period prior to interlock use. In all of these studies, crash rates are expressed in terms of time, rather than exposure. Only one study has included random assignment of offenders to interlock use or the control group, and it did not track crash rates. In all of the other analyses, offenders, judges, or hearing officers decide who receives an interlock, which results in a substantial selection-bias problem. Individuals who need to drive a lot and can afford to do so get interlocks; those who are poor or do not need to drive much accept license revocation. The latter often continue to do some driving, and they continue to be apprehended for DUI. Their annual VMT is thought to be much less than before their revocations, but data is lacking.

Consider the feasibility of alternative ways to monitor alcohol-impaired driving

- For example, low-cost video recording triggered by high-acceleration maneuvers might be used to obtain images of the road and the driver's face, which could be transmitted to an authority such as an employer or parent. Video summaries of the high acceleration events could be downloaded and replayed to see what was going on in the car and on the road, question the driver, and apply appropriate feedback.

Address the operational paradigms for impaired-driving detectors

- Who should impaired-driving detectors warn? Should they warn in real-time: the driver, the passengers, the surrounding motorists, vehicle-owner/parents via cell phone, authorities via cell phone?

- To whom should impaired-driving detectors report data? Should they report after-the-fact to vehicle owner/parents, authorities, and the insurance carrier?

- What restrictions should be placed on the vehicle's operation? Should vehicle use be restricted by preventing engine start, allow engine start for heat/air conditioning, but lock gearshift in park/neutral, limit top speed, limit top speed to a value appropriate to the road type using GPS?

- What is the potential consumer acceptance of the "Big Brother" aspects of impairment monitoring technologies? Is there a difference in acceptance of vehicle, as opposed to driver surveillance?

- Determine which vehicle actions should be constrained if impairment is detected. Is it useful to an impaired driver to reduce distraction such as turning off the radio, prompt the occupants to fasten seatbelts and check that the headlights are on?

Recognize that coordination and consistency in the administrative and legal processes supporting impairment detection will increase use of BAIIDs:

- Support consistent record keeping practices to facilitate the acquisition of information about recidivism to know if programs in place are working.
- Draft model legislation and guidelines for State programs and laws, as well as set specifications for hardware. Greater uniformity will allow BAIID vendors to serve the market more efficiently.
- Consider strategies to reduce the time interval between incident and DUI sanction because shorter time intervals are related to a lower incidence of recidivism. For example, police officers choose whether to use breath tests or blood and urine tests, and it can take up to 40 to 50 days to obtain the results of the latter.
- Identify loopholes in State laws that permit convicted DUI offenders easy access to alternative vehicles. For example, some States do not require the return of license plates upon the sale of a vehicle. During an interview it was mentioned that DUI offenders purchase inexpensive used cars because they come with license plates, thus making it possible to continue driving.
- People with many DUI convictions should be continuously monitored. Having IIDs in cars only partially helps people with chronic problems.
- Document the reasons for the 2.5-to-1 difference between States in the incidence of injury and fatality to due to alcohol-related accidents.
- Quantify how the level of law enforcement relates to the incidence of failed BAC tests.
- Identify how alcohol interacts with other impairment factors. Low BAC can interact with distraction for example, to produce effects comparable to BAC = .08, since alcohol constrains the capacity to multitask, and driving requires multitasking.

Reexamine the automotive infrastructure to enhance coordination with impairment monitoring

- Evaluate the development of standards for a secondary interlock interface that would reduce installation/removal costs and make circumvention difficult.
- Explore the potential for adding GPS capability to a technology to monitor and control impaired driving.
- Consider the possibility of bundling impaired-driving-prevention technologies with related technologies and services, such as navigation services, insurance discounts, roadside assistance, anti-theft, etc., to enhance consumer acceptance.

11. RESEARCH RECOMMENDATIONS FOR TECHNOLOGIES UNDER DEVELOPMENT

11.1 NIR Tissue Spectroscopy

Near term, establish the credibility of NIR -tissue spectroscopy for wider application in interlocks. Develop a rugged, battery-powered, suitcase NIR reflectance spectroscope for evidential use, which has the following advantages:

- As accurate as a Breathalyzer – even spectroscopic types;
- Eliminates 15-minute waiting period for clearance of mouth alcohol, thus improving throughput at checkpoints;
- Easier to use with uncooperative or unconscious subjects; and
- Renders moot defense arguments about contaminants and partition ratios.

Longer term, use of tissue spectroscopy requires the resolution of the following physiological issues:

- The soft, thin skin on the underside of the forearm works well for reflectance spectroscopy;
- Little is known about the reflectance characteristics of the thicker, tougher skin of the palms and fingers;
- Little is known about perfusion rates in various parts of the hand;
- Little is known about the effects of the bony structures that lie close to the skin;
- Individual variations caused by manual labor are likely to be large

A series of imaging studies using ultrasound and infrared technologies is proposed to explore these questions. Tests must include a variety of subjects representative of the full range of variability in the relevant physiological characteristics. This data will determine the strength and quality of the reflected signal at various locations on the hand. This will establish which, if any, points on the hand are usable targets and set performance requirements for the sensor.

11.2 Role of Warning Devices

Evaluate the efficacy of impairment warnings and incentive displays by establishing a general baseline definition of impairment that is based upon crash risk and use it in assessing the effectiveness of impairment warnings, incentives, and crash avoidance warnings.

Research is needed to define ways to display warning and incentive information that will not increase impairment through distraction. One specific example is the current lane departure warning system. If alcohol impaired drivers focus their attention on the road center, then the warning system may need to be displayed within their reduced visual field if it is to be effective in preventing alcohol impairment crashes.

- Finer-grained data is needed to untangle the comorbidity between alcohol and drugs, clarify the role of DUI offenders (first-time and recidivists) and disaggregate risk factors by vehicle type in the causation of all crashes, including fatalities.
- Track the new European program DRiving Under the Influence of Drugs (DRUID) to understand whether it will attempt to discriminate alcohol effects from those of drugs and medicines.
- Conduct much of any impaired driving research at night. Document how often fatigue and darkness are factors in the crashes of alcohol-abusing drivers. Most studies on illegal drug effects in driving have been done in daytime, while these drugs are mostly taken in the evening and at night. For that reason, the drug effects should be assessed in combination with sleep deprivation. Only then can their "real" effects on traffic safety be estimated.
- Develop a general set of vehicle sensor thresholds that relate directly to risk and apply to everyone. Weigh the advantages and disadvantages of personal versus general thresholds for impairment detection. Consider ways of combining them.
- Assess the attitudes of drivers toward various forms of feedback about driving performance that might be generated by emerging technologies.
- Assess the attitudes of passengers to impairment warning and of drivers and passengers to warnings that alcohol has been detected in the vehicle.

11.3 A GENERAL BEHAVIORAL BASELINE FOR IMPAIRMENT DETECTION

Perform research to produce a general risk-based baseline for behavioral impairment detection.

The objective of this work is to determine the threshold values for driving-behavior sensors to detect impairment. These thresholds would then be available for use in a general risk-based baseline definition of impairment applicable to all drivers.

11.3.1 Background and Research Strategy

There are two potential research strategies to create a general risk-based baseline for impairment detection:

1. Direct experimental determination of crash risk; and
2. Statistical determination of crash risk.

According to DeGier[74] and Ramaekers[75] it is impractical to establish crash risk thresholds directly, and so they recommended determining the threshold indirectly as the effect of BAC at the per se limit on the metrics of interest. Since the relation of BAC to crash risk is well accepted, thresholds found in this way would be indirectly linked to crash risk. For example, Louwerens et al.[35] determined the relationship between BAC and two vehicle performance parameters: lane position variability and speed variability. These results were not replicated[106], reflecting the difficulty and risks of this approach.

The limited accuracy of Project SAVE impairment detection may have been caused in part by the subjective definition of "impairment" that was used to train the neural network on the distinction between impaired and unimpaired driving performance (a driving instructor provided performance and fatigue ratings that were later used to define measures of impairment). A distinction should be made between the accuracy required in a warning system versus that required for an interlock. Accuracies on the order of 70 percent may be acceptable for warning systems.

Additional aspects of the SAVE evaluation that could have reduced its apparent effectiveness include:

- Detection of impairment at the BAC = .05 per se limit, "which is a very mild rate, hence they do not represent serious impairment and thus the relevant detection is rather low" ([88], p. 13);
- Small sample size of nine or fewer subjects; and
- Lack of combined fatigue and alcohol conditions.

DeGier and Ramaekers did not describe their reasons for considering the direct determination of crash risk to be problematic, but an obvious difficulty is that few if any crashes can be expected under typical conditions in a simulation experiment, even when the subjects are impaired. Direct experimental determination of crash risk would require a departure from these conditions to increase crash risk and yet represent real world conditions.

Suggestions to realistically increase simulation crash rate include:

- *Subject selection:* Provide a representative sample of adults who consume alcohol except that the study must exclude individuals under treatment for alcohol abuse and chronic alcoholics for ethical reasons. They should sample both the range in duration of drinking experience and current drinking behavior (light, moderate, or heavy). Demographic characteristics that focus on the populations most prone to alcohol impaired driving (e.g., age, gender, ethnicity) will help to provide results that will generalize to the population of alcohol impaired drivers.
- *Research Facilities*: The study of one particularly important dependent variable, vehicle speed when negotiating a curve, appears to require

closed course facilities. Gawron and Ranney[107] found that drivers impaired due to BAC = .12 failed to reduce vehicle speed prior to curves, while at BAC = .07 or less, drivers did reduce speed. The authors suggest that this difference may have resulted from the absence of lateral acceleration cues in the simulation facility.

- *Simulation scenarios*: Test track and simulation scenarios should be designed to address the situations in which alcohol-related crashes are most common. Four times as many alcohol-related crashes occur during the night as during the day.[2] GES data indicates that in addition to night driving, inclement weather (wet road surface, poor visibility) is also associated with alcohol-related crashes (Table 11-1). While most occur on straight road segments, alcohol-involved drivers are overrepresented in crashes on curves.[33] The Volpe Center interview with Lt. R. Reichert of the Washington State Patrol found that to detect impaired driving, officers look for wide turns at intersections to the point of crossing into the adjacent lane, weaving back and forth across lanes, driving without headlights at night, failure to use a turn signal, jerky steering actions, or collisions. Also consider findings from laboratory and vehicle research to identify situations that require divided attention between the road ahead and events at roadside.

- *Blood Alcohol:* BAC should represent the alcohol concentrations found in fatal crash statistics as well as lower amounts. The blood alcohol analysis of drivers involved in fatal crashes in 2004 indicates an average BAC level of .16 and a 75th percentile BAC of .20.[108] BAC should be manipulated through ad libitum administration with instructions to consume a normal amount, but no more than they would before driving. In no case should a dose be provided that would result in a medically dangerous BAC.

- *Baseline Conditions:* Activities such as operating a vehicle sound system or conversing with a passenger that are considered safe by consensus are needed for baseline conditions to ensure that the baseline will not identify consensually safe activities as impaired when tested under placebo. They should be implemented in a way that provides the subject with the same latitude to engage in the activity (or not) as occurs under naturalistic conditions.

Table 11-1 Alcohol involvement in crash scenarios

	Alcohol - Yes	Alcohol - No	Yes-to-No Ratio
Single-Vehicle Crashes			
Road edge departure/no maneuver Vehicle is going straight in a rural area at night, under clear weather, with a posted speed limit ≥ 55 mph, and then departs the edge of the road at a non-junction area.	93,350	236,706	.39
Control loss/no prior vehicle action Vehicle is going straight in a rural area, in daylight, under adverse weather conditions, with a posted speed limit ≥ 55 mph, and then loses control due to wet/slippery roads and runs off the road.	60,347	410,492	.15
Road edge departure/maneuver	14,021	51,520	.27
Two-Vehicle Crashes			
Rear-end/LVS	28,719	735,960	.04
Opposite direction/no maneuver	9,876	163,292	.06
Running red light	9,803	419,391	.06

11.3.2 Risks

Variability in the effects of alcohol and in the ability to compensate for its effects are challenges to accomplishing the goals of this research. These risks may be mitigated by the inclusion of higher BAC conditions, which tend to result in greater consistency.[109]

There are also inherent risks associated with alcohol consumption, for example, by individuals with unknown health conditions. A thorough medical screening of subjects is essential.

11.4 COUNTERMEASURE RESEARCH

11.4.1 Impairment and Incentive Displays

Perform studies to evaluate impairment warnings and incentive displays.
After establishing a general baseline definition of impairment, the effectiveness of impairment warnings, incentives, and crash avoidance warnings can be assessed to determine their effects on impaired driving. Research will be needed on specific ways to display warning and incentive information that will not increase impairment through distraction. One specific example is the current lane departure warning system. If alcohol-impaired drivers focus their attention on the road center, then the warning system may need to be displayed within their reduced visual field if it is to be effective in preventing alcohol-impairment crashes. An alternative hypothesis is that impaired drivers will increase their visual field to incorporate valid and well-designed displays. The effectiveness of aural, haptic, and multi-modal displays should also be assessed.

12. REFERENCES

1. Medford, R., *A Nation Without Drunk Driving*, in *International DUI Technology Symposium*. 2006: Albuquerque.
2. *Traffic Safety Facts 2004: A Compilation of Motor Vehicle Crash Data from the Fatality Analysis Reporting System and the General Estimates System*. 2005. National Center for Statistics and Analysis. Washington, DC: National Highway Traffic Safety Administration.
3. Peters, B., & van Winsum, W., *SAVE: System for Effective Assessment of the Driver State and Vehicle Control in Emergency Situations*. 1998. R&D Program Telematics: EU.
4. Liu, C., Chen, C. S., Subramanian, R., & Utter, D., *Analysis of Speeding-Related Fatal Motor Vehicle Crashes*. 2005. Washington, DC: National Highway Traffic Safety Administration.
5. Fell, J. C. *Potential Role of Technology in Reducing Alcohol-Related Traffic Fatalities*, in *International Technology Symposium: A Nation Without Drunk Driving*. 2006. Albuquerque, New Mexico: MADD.
6. Grant, B. F., & Dawson, D. A., *Introduction to the National Epidemiologic Survey on Alcohol and Related Conditions*, National Institute on Alcohol Abuse and Alcoholism.
7. NCSA, *Traffic Safety Facts-Crash Stats*. 2006. National Center for Statistics and Analysis. Washington, DC: National Highway Traffic Safety Administration.
8. NHTSA, *Analysis of Speeding-Related Fatal Motor Vehicle Crashes*. 2005. Washington, DC: National Highway Traffic Safety Administration.
9. Fell, J. C., *Potential Role of Technology in Reducing Alcohol-Related Traffic Fatalities*, in *International Technology Symposium: A Nation Without Drunk Driving*. 2006, Albuquerque, NM: MADD
10. *Crime in the United States 2004: Uniform Crime Reports*. 2005. Washington, DC: Federal Bureau of Investigation .
11. NHTSA, *National Survey of Drinking and Driving Attitudes and Behaviors, 2001*. Traffic Safety Facts, 2003. **280**. Washington, DC: National Highway Traffic Safety Administration.
12. Balmforth, D., *National Survey of Drinking and Driving Attitudes and Behavior: 1997*. 1999. NHTSA and the Gallup Organization.HS-808 844. Washington, DC: National Highway Traffic Safety Administration.
13. Royal, D., *National Survey of Drinking and Driving Attitudes and Behaviors, 2001*, in *Traffic Safety Facts*, 2003. NHTSA and the Gallup Organization. DOT-HS-809-549. Washington, DC: National Highway Traffic Safety Administration.
14. Williams, A. F., *Alcohol-Impaired Driving and its Consequences in the United States: The Past 25 Years*. Journal of Safety Research, 2006. **37**(2): p. 123-38.
15. Beirness, D. J., Mayhew, D. R., & Simpson, H. M., *Highlights DWI Repeat Offenders: A Review and Synthesis of the Literature* 1997. Health Canada.
16. Harrison, E. L. R., & Fillmore, M. T., *Are Bad Drivers More Impaired by Alcohol? Sober Driving Precision Predicts Impairment from Alcohol in a Simulated Driving Task*. Accident Analysis & Prevention, 2005. **37**: p. 882-889.
17. NHTSA, *Traffic Safety Facts: Repeat Intoxicated Driver Laws*. 2004. Washington, DC: National Highway Traffic Safety Administration.
18. Lapham, S. C., Kapitula, L. R., C'de Baca, J., & McMillan, G. P., *Impaired-Driving Recidivism Among Repeat Offenders Following an Intensive Court-Based Intervention*. Accident Analysis & Prevention, 2006. **38**: p. 162-169.

19. Frost, C. J., Phillips, M. E., Tollefson, D., & Werstak, J., *What We Know About Offenders Who Drive Under the Influence: Analysis of Court Case File Reviews*. Accident Analysis & Prevention, 2006. **38**: p. 84-91.
20. Nochajski, T. H. & Stasiewicz, P. R., *Relapse to Driving Under the Influence (DUI): A Review*. Clinical Psychology Review, 2006. **26**: p. 179-195.
21. Volpe National Transportation Systems Center, *Technology to Prevent Impairment Crashes: Interview Summaries*. 2006. Cambridge, MA: Department of Transportation.
22. Substance Abuse and Mental Health Services Administration (SAMSA), *Binge Drinking Among Underage Persons*. 2002. Washington, DC: Department of Health and Human Services.
23. Office of Alcohol and Other Drug Abuse, *Binge Drinking on the Rise in U.S.*, in *American Medical Association*. 2003.
24. Centers for Disease Control, *Alcohol Use Among Adolescents and Adults --- New Hampshire, 1991--2003*. Morbidity and Mortality Weekly Report, 2004. **53**((08)): p. 174-175.
25. Substance Abuse and Mental Health Administration, *Binge Drinking Among Underage Persons*. 2002, The National Household Survey on Drug Abuse.
26. NHTSA, *Initiatives to Address Impaired Driving*. 2003. Washington, DC: National Highway Traffic Safety Administration.
27. Subramanian, R., *Total and Alcohol-Related Fatality Rates by State, 2003-2004*. 2006. National Center for Statistics and Analysis. Washington, DC: National Highway Traffic Safety Administration.
28. NHTSA, *National Survey of Drinking and Driving Attitudes and Behaviors, 2001*. 2003. Washington, DC: National Highway Traffic Safety Administration.
29. National Center for Statistics and Analysis, *Traffic Safety Facts State Alcohol Estimates*. Washington, DC: National Highway Traffic Safety Administration.
30. Federal Register, *Model Specifications for Breath Alcohol Ignition Interlock Devices*. 1992. p. 11774-11787.
31. Borkenstein, R. F., Crowther, R. F., Shumate, R. P., Zeil, W. B., & Zylman, R., *The Role of the Drinking Driver in Traffic Accidents*. 1964. p. 136-245.
32. Kerr, J. S. & Hindmarch, I., *The Effects of Alcohol Alone or in Combination with Other Drugs on Information Processing*. Human Psychopharmacology, 1998. **13**: p. 1-9.
33. Terhune, K. W., Ippolito, C. A., Hendricks, D. L., Michalovic, J. G., Bogema, S. C., Santinga, P., Blomberg, R., & Preusser, D. F., *The Incidence and Role of Drugs in Fatally Injured Drivers*. 1992. DOT HS 808 065. Washington, DC: National Highway Traffic Safety Administration.
34. Robbe, H., *Marijuana's Impairing Effects on Driving are Moderate When Taken Alone but Severe When Combined with Alcohol*. Human Psychopharmacology, 1998. **13**: p. S70-S78.
35. Louwerens, J. W., Gloerich, A. B. M., & Vries, G., *The Relationship Between Drivers' Blood Alcohol Concentration (BAC) and Actual Driving Performance During High Speed Travel*, in *Alcohol, Drugs and Traffic Safety-T86. Proceedings of the 10th International Conference on Alcohol, Drugs and Traffic Safety*, P.C. Noordzij & R. Roszbach, Editors. 1987. p. 183-186.
36. De Waard, D. & Brookhuis, K. A., *Assessing Driver Status: A Demonstration Experiment on the Road*. Accident Analysis & Prevention, 1991. **23**: p. 297-307.
37. Lamers, C. T. J., & Ramaekers, J. G., *Visual Search and Urban City Driving Under the Influence of Marijuana and Alcohol*. Human Psychopharmacology Clinical and Experimental, 2001. **16**: p. 393-401.
38. Keall, M. D., Frith, W. J., & Patterson, T. L., *The Contribution of Alcohol to Night Time Crash Risk and Other Risks of Night Driving*. Accident Analysis & Prevention, 2005. **37**: p. 816-824.

39. Keall, M. D., Frith, W. J., & Patterson, T. L., *The Influence of Alcohol, Age, and Number of Passengers on the Night Time Risk of Driver Fatal Injury in New Zealand.* Accident Analysis and Prevention, 2004. **36**(1): p. 49-61.
40. Banks, S., Catcheside, P., Lack, L., Grunstein, R. R., & McEvoy, R. D., *Low Levels of Alcohol Impair Driving Simulator Performance and Reduce Perception of Crash Risk in Partially Sleep Deprived Subjects.* Sleep, 2004. **27**(6): p. 1063-1067.
41. Horne, J. A., Reyner, L. A., & Barrett, P. R., *Driving Impairment Due to Sleepiness is Exacerbated by Low Alcohol Intake.* Occupational and Environmental Medicine, 2003. **60**(9): p. 689-693.
42. Barrett, P. R., Horne, J. A., & Reyner, L. A., *Sleepiness Combined with Low Alcohol Intake in Women Drivers: Greater Impairment but Better Perception than Men?* Sleep, 2004. **27**(6): p. 1057-1062.
43. Wang, M. Q., Nicholson, M. E., Mahoney, B. S., Li, L., & Perko, M. A., *Proprioceptive Responses Under Rising and Falling BACs: A Test of the Mellanby Effect.* Perceptual and Motor Skills, 1993. **77**: p. 83-88.
44. Grattan-Miscio, K. E. & Vogel-Sprott, M., *Effects of Alcohol and Performance Incentives on Immediate Working Memory.* Psychopharmacology, 2005. **181**: p. 188-196.
45. Moskowski, H. & Robinson, C., *Effects of Low Doses of Alcohol on Driving-Related Skills: A Review of the Evidence.* 1988. DOT HS 807 280. Washington, DC: National Highway Traffic Safety Administration, SRA Technologies, Inc.
46. DeYoung, D. J., Tashima, H. N., & Masten, S. V., *An Evaluation of the Effectiveness of Ignition Interlock in California.* 2004. CAL-DMV-RSS-04-210. Sacramento, CA: California Department of Motor Vehicles.
47. DeYoung, D. J., Tashima, H. N., & Masten, S. V. *An Evaluation of the Effectiveness of Ignition Interlock in California.* in *Alcohol Ignition Interlock Devices Volume II: Research, Policy, and Program Status.* 2005. Sacramento, CA: California Department of Motor Vehicles.
48. Vézina, L. *The Québec Alcohol Ignition Interlock Program: Impact on Recidivism and Crashes.* in *16th International Conference on Alcohol, Drugs and Traffic Safety.* 2002. Québec City: Société de l'assurance automobile du Québec.
49. Morse, B. J., & Elliot, D. S., *Effects of Ignition Interlock Devices on DWI Recidivism: Findings From a Longitudinal Study in Hamilton County, OH.* Crime & Delinquency, 1992. **38**(2): p. 131-157.
50. Voas, R., Roth, R., & Marques, P. *Interlocks for First Offenders: Effective?* in *Sixth International Symposium on Alcohol Ignition Interlock Programs.* 2005. Annecy, France.
51. Kizev, W. & Turner, A., *Wade Kizev, Commonwealth's Attorney for Henrico County, Virginia and Amy Turner, Assistant Commonwealth's Attorney (Assigned to Traffic Court).* 2006.
52. Ridder, T. D., Brown, C. D., & Ver Steeg, B. G., *Framework for Multivariate Selectivity Analysis, Part II: Experimental Applications.* Applied Spectroscopy, 2005. **59**(6): p. 804-815.
53. Ridder, T. D., Hendee, S. P., & Brown, C. D., *Noninvasive Alcohol Testing Using Diffuse Reflectance Near-Infrared Spectroscopy.* Applied Spectroscopy, 2005. **59**(2): p. 181-189.
54. Ghionea, S., *Ethanol Sensing for Detecting Blood Alcohol Concentration.* 2006, Unpublished Manuscript, Oregon State University. Available from the authors.
55. McCain, S. T. e. a., Gehm, M. E., Wang, Y., Pitsianis, N. P., & Brady, D. J., *Coded-aperture Raman Spectroscopy for Quantitative Measurements of Ethanol in a Tissue Phantom.* Applied Spectroscopy, 2006. **60**: p. 663-71.
56. McCain, S. T., *Coded Spectroscopy for Ethanol Detection in Diffuse, Florescent Media*, in *Department of Electrical and Computer Engineering.* 2007, Duke University.
57. Backhouse, C., *Interview With Professor Chris Backhouse.* 2006.

58. Ennis, M., *Interview with Matthew Ennis, Lumidigm, Inc.* 2006.
59. Malyugin, S., Moskalev, T., Nadezhdinskii, A., Namestnikov, D., Ponurovskii, Y., Shapovalov, Y., Stavrovskii, D., Vyazov, I., Zaslavskii, V., & Upendra, R. *Ethanol Vapor Detection Limited by Diode Laser Frequency Quantum Noise.* in *5th International Conference on Tunable Diode Laser Spectroscopy.* 2005. Florence, Italy
60. Penza, M., Cassano, G., Adversa, P., Antolini, F., Cusano, A., Cutolo, A., Giordano, M., & Noicolais, L., *Alcohol Detection Using Carbon Nanotubes Acoustic and Optical Sensors.* Applied Physics Letters, 2004. **85**(12): p. 2379-2381.
61. Toan, N. N., Giang, H. T., Hieu, N. S., Thu, D. T. A., & Phuc, N. X. *Fabrication of Ethanol Detector on Basic of Nanosize Perovskite Oxides.* in *Eighth German-Vietnamese Seminar on Physics and Engineering.* 2005. Erlangen, Germany.
62. Swette, L. L., Griffith, A. E., & LaConti, A. B., *Potential and Diffusion Controlled Solid Electrolyte Sensor for Continuous Measurement of Very Low Levels of Transdermal Alcohol.* 1999, Giner, Inc.: United States.
63. Marques, P. & McKnight, S. *Evaluation of Transdermal Alcohol Devices.* in *6th Annual Innovative Technologies for Community Corrections Conference.* 2006. Seattle, Washington.
64. Stapleton, J. M., Guthrie, S., & Linnoila, M., *Effects of Alcohol and Other Psychotropic Drugs on Eye Movements: Relevance to Traffic Safety.* Journal of Studies on Alcohol, 1986. **47**(5): p. 426-432.
65. Nawrot, M., Nordenstrom, B., & Olson, A., *Disruption of Eye Movements by Ethanol Intoxication Affects Perception of Depth From Motion Parallax.* Psychological Science, 2004. **15**(12): p. 858-865.
66. Kosnowski, E. M., Yolton, R. L., Citek, K., Hayes, C. E., & Evans, R. B., *The Drug Evaluation Classification program: Using Ocular and Other Signs to Detect Drug Intoxication.* Journal of the American Optometric Association, 1998. **69**(4): p. 211-227.
67. Victor, T. W., Harbluk, J. L., & Engstrom, J. A., *Sensitivity of Eye-Movement Measures to In-Vehicle Task Difficulty.* Transportation Research Part F, 2005. **8**: p. 167-190.
68. Brookhuis, K., *How to Measure Driving Ability Under the Influence of Alcohol and Drugs, and Why.* Human Psychophamacology: Clinical and Experimental, 1998. **13**: p. S64-S69.
69. Advanced Brain Monitoring Inc. *EEG Technology.* 2006 [cited 2006 July 28]; Available from: http://www.b-alert.com/EEG.html.
70. Izzetoglu, K., Bunce, S., Onaral, B., Pourrezaei, K., & Chance, N., *Functional Optical Brain Imaging Using Near-Infrared During Cognitive Tasks.* International Journal of Human-Computer Interaction, 2004. **17**(2): p. 211-227.
71. St. John, M., Kobus, D. A., Morrison, J. G., & Schmorrow, D., *Overview of the DARPA Augmented Cognition Technical Integration Experiment.* International Journal of Human-Computer Interaction, 2004. **17**(2): p. 131-149.
72. Russo, M. B., Stetz, M. C., & Thomas, M. L., *Monitoring and Predicting Cognitive State and Performance via Psychological Correlates of Neuronal Signals.* Aviation, Space, and Environmental Medicine, 2005. **76**(7): p. C59-C63.
73. Moskowitz, H., *Attention Tasks as Skills Performance Measures of Drug Effects.* British Journal of Clinical Pharmacology, 1984. **18**: p. 51S-61S.
74. De Gier, J. J., *Drugs and Driving Research: Application of Results by Drug Regulatory Authorities.* Human Psychopharmacology: Clinical & Experimental, 1998. **13**: p. 133-136.
75. Ramaekers, J. G., *Antidepressants and Driver Impairment: Empirical Evidence From a Standard On-the-Road Test.* Journal of Clinical Psychiatry, 2003. **64**(1): p. 20-29.

76. Burns, P. C., Parkes, A., Burton, S., Smith, R. K., & Burch, D., *How Dangerous Is Driving With a Mobile Phone? Benchmarking the Impairment to Alcohol.* 2002. Transport Research Laboratory and Direct Line Group.
77. Strayer, D. L., Drews, F. A., & Crouch, D. J., *A Comparison of the Cell Phone Driver and the Drunk Driver.* Human Factors, 2006. **48**(2): p. 381-391.
78. Tiplady, B., Hiroz, J., Holmes, L., & Drummond, G., *Errors in Performance Testing: A Comparison of Ethanol and Temazepam.* Journal of Psychopharmacology, 2003. **17**(1): p. 41-49.
79. Mills, K. C., Spruill, S. E., Kanne, R. W., Parkman, K. M., & Zhang, Y., *The Influence of Stimulants, Sedatives, and Fatigue on Tunnel Vision: Risk Factors for Driving and Piloting.* Human Factors, 2001. **43**(2): p. 310-327.
80. Krüger, H.-P. & Berghaus, G. *Behavioral Effects of Alcohol and Cannabis: Can Equipotencies be Established?* in *13th International Conference on Alcohol, Drugs and Traffic Safety.* 1995. Adelaide, Australia.
81. Harrison, E. L. R. & Fillmore, M. T., *Social Drinkers Underestimate the Additive Impairing Effects of Alcohol and Visual Degradation on Behavioral Functioning.* Psychopharmacology, 2005. **177**: p. 459-464.
82. Williamson, A., Feyer, A.-M., Friswell, R., & Finlay-Brown, S., *Development of Measures of Fatigue: Using an Alcohol Comparison to Validate the Effects of Fatigue on Performance.* 2000. School of Psychology and NSW Injury Risk Management Research Centre, University of New South Wales, New Zealand Occupational and Environmental Health Research Centre, Australian Transport Safety Bureau.
83. Arnedt, J. T., Wilde, G. J. S., Munt, P. W., & MacLean, A. W., *How do Prolonged Wakefulness and Alcohol Compare in the Decrements they Produce on a Simulated Driving Task?* Accident Analysis & Prevention, 2001. **33**: p. 337-344.
84. Langer, P., Holzner, B., Magnet, W., & Kopp, M., *Hands-Free Mobile Phone Conversation Impairs the Peripheral Visual System to an Extent Comparable to an Alcohol Level of 4-5 g 100 ml.* Human Psychopharmacology: Clinical & Experimental, 2005. **20**: p. 65-66.
85. Tiplady, B., Degia, A., & Dixon, P., *Assessment of Driver Impairment: Evaluation of a Two-Choice Tester Using Ethanol.* Transportation Research Part F, 2005. **8**: p. 299-310.
86. Brookhuis, K. A., de Waard, D., & Fairclough, S. H., *Criteria for Driver Impairment.* Ergonomics, 2003. **46**(5): p. 433-45.
87. Ward, N. J., *Personal Communication.*
88. Bekiaris, E., *System for Effective Assessment of the Driver State and Vehicle Control in Emergency Situations: Final Report.* 1999. Reference No. TR 1047, D.20. R&D Telematics: EU.
89. Bekiaris, E., Nikolaou, S., & Tzovaras, D. *Sensors for Unobtrusive Driver Fatigue Monitoring: from AWAKE to SENSATION.* in *Intelligent Transportation Society of America Conference Proceedings.* 2005. Washington, DC: ITS World Congress.
90. Nau, P. A., Van Houten, R., Rolider, A., & Jonah, B. A., *The Failure of Feedback on Alcohol Impairment to Reduce Impaired Driving.* Journal of Applied Behavior Analysis, 1993. **26**: p. 361-367.
91. Fairclough, S. H. & van Winsum, W., *The Influence of Impairment Feedback on Driver Behavior: A Simulator Study.* Transportation Human Factors, 2000. **2**(3): p. 229-246.
92. Pilcher, J. J., Nadler, E., & Busch, C., *Effects of Hot and Cold Temperature Exposure on Performance: A Meta-Analytic Review.* Ergonomics, 2002. **45**(10): p. 682-698.
93. Holloway, F. A., *Low-Dose Alcohol Effects on Human Behavior and Performance.* Alcohol, Drugs and Driving, 1995. **11**(1): p. 39-56.

94. Grattan-Miscio, K. E., & Vogel-Sprott, M., *Alcohol, Intentional Control, and Inappropriate Behavior: Regulation by Caffeine or an Incentive.* Experimental and Clinical Psychopharmacology, 2005. **13**(1): p. 48-55.
95. Fillmore, M. T., *Environmental Dependence of Behavioral Control Mechanisms: Effects of Alcohol and Information Processing Demands.* Experimental and Clinical Psychopharmacology, 2004. **12**(3): p. 216-223.
96. Iudice, A., Bonanni, E., Gelli, A., Frittelli, C., Iudice, G., Cicgnoni, F., Ghicopulos, I., & Muini, I., *Effects of Prolonged Wakefulness Combined with Alcohol and Hands-Free Cell Phone Divided Attention Tasks on Simulated Driving.* Human Psychopharmacology: Clinical & Experimental, 2005. **20**: p. 125-132.
97. Rakauskas, M. & Ward, N. *Behavioral Effects of Driver Distraction and Alcohol Impairment.* in *The 49th Annual Meeting of the Human Factors and Ergonomics Society.* 2005. Orlando, Florida.
98. Rakauskas, M. E., Ward, N. J., Bernat, E., Cadwallader, M., & Patrick, C., *Car Following Performance During Conventional Distractions and Alcohol Intoxication.* Human Factors, Submitted for review April, 2006.
99. Recarte, M. A. & Nunes, L. M., *Effects of Verbal and Spatial-Imagery Tasks on Eye Fixations While Driving.* Journal of Experimental Psychology: Applied, 2000. **6**(1): p. 31-43.
100. Recarte, M. A. & Nunes, L. M., *Mental Workload While Driving: Effects on Visual Search, Discrimination, and Decision Making.* Journal of Experimental Psychology: Applied, 2003. **9**(2): p. 119-137.
101. Moskowitz, H., *Alcohol and Drugs*, in *Human Factors in Traffic Safety*, R.E. Dewar and P.L. Olson, Editors. 2002, Lawyers and Judges Publishing Company: Tucson. p. 177-207.
102. Roetting, M., Huang, Y.-H., McDevitt, J. R., & Melton, D., *When Technology Tells You How You Drive - Truck Drivers' Attitudes Towards Feedback by Technology.* Transportation Research Part F, 2003. **6**: p. 275-287.
103. Huang, Y.-H., Roetting, M., McDevitt, J. R., Melton, D., & Smith, G. S., *Feedback by Technology: Attitudes and Opinions of Truck Drivers.* Transportation Research Part F, 2005. **8**: p. 277-297.
104. NHTSA, *Impaired Driving in the United States.* 2000. Washington, DC: National Highway Traffic Safety Administration.
105. Harless, D. W., & Hoffer, G. E., *The Antilock Braking System Anomaly: A Drinking Driver Problem?* Accident Analysis & Prevention, 2002. **34**: p. 333-341.
106. De Waard, D., & Brookhuis, K. A., *Assessing Driver Status: A Demonstration Experiment on the Road.* Accident Analysis and Prevention, 1991. **23**(4): p. 297-307.
107. Gawron, V. J., & Ranney, T. A., *The Effects of Alcohol Dosing on Driving Performance on a Closed Course and in a Driving Simulator.* Ergonomics, 1988. **31**(9): p. 1219-1244.
108. Subramanian, R., *Alcohol Involvement in Fatal Motor Vehicle Crashes, 2005.* 2005. Washington, DC: National Highway Traffic Safety Administration.
109. Moskowitz, H., & Fiorentino, D., *A Review of the Literature on the Effects of Low Doses of Alcohol on Driving-Related Skills.* 2000. Washington, DC: National Highway Traffic Safety Administration.

DOT HS 810 833
September 2007

www.ingramcontent.com/pod-product-compliance
Lightning Source LLC
Chambersburg PA
CBHW080304180526
45167CB00006B/2669